ANCIENT LANDSCAPES
OF THE COLORADO PLATEAU

Ron Blakey and Wayne Ranney

GRAND CANYON ASSOCIATION

Grand Canyon Association
P.O. Box 399
Grand Canyon, AZ 86023-0399
(800) 858-2808
www.grandcanyon.org

Printed in China on recycled paper using soy-based inks
Edited by Pam Frazier and Todd Berger
Designed by Ron Short

14   13   12   4   5   6

Library of Congress Cataloging-in-Publication Data

Blakey, Ronald C., 1945–
 Ancient landscapes of the Colorado plateau / Ronald C. Blakey and
Wayne Ranney. -- 1st ed.
     p. cm.
 Includes bibliographical references and index.
 ISBN-13: 978-1-934656-03-7 (alk. paper)
 1. Geology, Structural--Colorado Plateau. 2. Paleogeography--Colorado
Plateau. 3. Geological time--Colorado Plateau. 4. Landscape
changes--Colorado Plateau--History. 5. Digital mapping--Colorado
Plateau. I. Ranney, Wayne. II. Title.
   QE627.5.C6.B63 2008
   557.88'1--dc22

                    2008006875

Additional Photograph and Illustration Credits:

Ron Blakey: xiv, 4, 8, 10 (top), 18, 19, 21 (bottom), 27 (bottom), 31 (bottom), 34 (bottom), 40 (top), 40 (bottom), 41 (top), 44 (bottom), 50, 52 (bottom), 54, 58 (top), 58 (middle), 58 (bottom), 59 (top), 59 (bottom), 65 (bottom), 66 (bottom), 67 (top), 72 (top), 73 (top). 78 (background), 79 (top), 79 (bottom), 84, 85 (top), 85 (bottom), 88 (bottom), 93 (top), 93 (bottom), 100 (top), 101 (top), 110–111, 114, 115, 116 (top), 116 (bottom), 127, 128 (bottom), 130 (top), 130 (bottom), 134 (bottom), 135 (top), 136 (bottom), 138 (bottom), 139 (bottom), 141 (top), 141 (bottom), 142

Dee Blakey: back flap

Chalk Butte, Inc., 126

Wayne Ranney: xv, 2, 5 (top), 10 (bottom), 11 (bottom), 14 (bottom), 21 (top), 22 (bottom), 26 (top), 26 (bottom), 27 (top), 31 (top), 35 (top), 51, 52 (top), 55, 64 (top), 64 (bottom), 65 (top), 66 (top), 67 (bottom), 72 (bottom), 73 (bottom), 92 (top), 92 (bottom), 100 (bottom), 104 (top), 105 (bottom), 110 (top), 110 (bottom), 111 (top), 118, 119 (top), 119 (bottom), 120, 128 (top), 132 (top), 132 (bottom), 133 (top), 133 (bottom), 135 (bottom), 137, 138 (top), 144, 145 (top), front flap

National Park Service, 124–125

Ron Short: 5 (bottom), 11 (top)

William K. Hartmann, 101 (bottom)

Wood Ronsaville Harlin, Inc.: 20 (top), 31 (bottom), 33 (bottom)

It is the mission of the Grand Canyon Association to cultivate knowledge, discovery, and stewardship for the benefit of Grand Canyon National Park

and its visitors. Proceeds from the sale of this book will be used to support the educational goals of Grand Canyon National Park.

DEDICATIONS

This book is dedicated to my family, Dee, Josh, and Ian, who accompanied me into the field and sometimes saw more rocks than they bargained for; to my thesis advisors, the late William Lee Stokes (University of Utah) and the late Bill Furnish (University of Iowa), who helped me read the rocks; and to Northern Arizona University's geology graduate students who gathered much of the data on which this book is based. —Ron Blakey

This book is dedicated to my father, Don Ranney, who, as a surveyor, collected fossils from the hills in Southern California and brought them home to his three sons and daughter. This single act of thoughtful kindness fueled my love for landscapes and the maps that took me there. Thank you, Dad! —Wayne Ranney

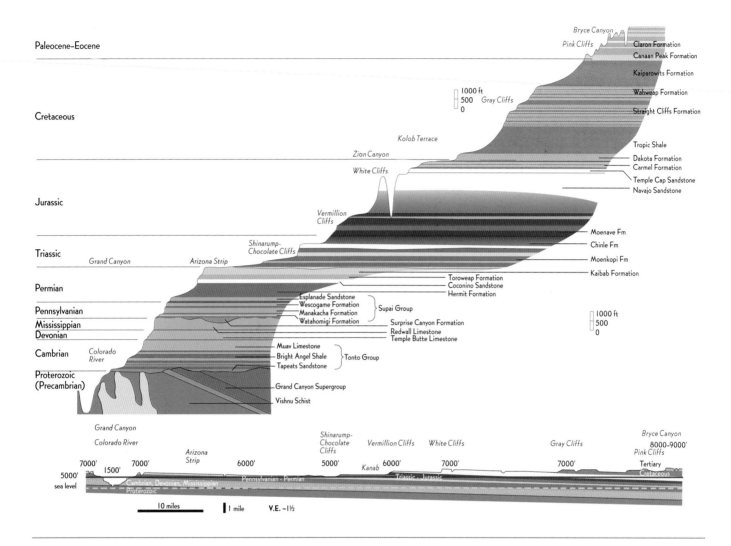

Paleocene-Eocene

Bryce Canyon
Pink Cliffs — Claron Formation
Canaan Peak Formation
Kaiparowits Formation

Cretaceous
1000 ft
500
0 — Gray Cliffs — Wahweap Formation
Straight Cliffs Formation

Kolob Terrace
Tropic Shale
Zion Canyon — Dakota Formation
White Cliffs — Carmel Formation
Temple Cap Sandstone
Navajo Sandstone

Jurassic

Vermillion
Cliffs

Triassic
Grand Canyon — Arizona Strip — Shinarump-
Chocolate Cliffs — Moenave Fm
Chinle Fm
Moenkopi Fm
Kaibab Formation

Permian
Toroweap Formation
Coconino Sandstone
Hermit Formation

Esplanade Sandstone
Wescogame Formation ⎫ Supai Group
Manakacha Formation ⎬
Pennsylvanian
Watahomigi Formation ⎭
Mississippian
Surprise Canyon Formation
Devonian
Redwall Limestone
Temple Butte Limestone

1000 ft
500
0

Cambrian
Colorado
River
Muav Limestone
Bright Angel Shale ⎫ Tonto Group
Tapeats Sandstone ⎬

**Proterozoic
(Precambrian)**
Grand Canyon Supergroup

Vishnu Schist

Grand Canyon
Colorado River

Shinarump-
Chocolate
Cliffs
Vermillion Cliffs — White Cliffs
Gray Cliffs
Bryce Canyon
8000-9000'
Pink Cliffs

7000'   7000'   Arizona
Strip   6000'   5000'   Kanab 6000'   7000'   7000'   Tertiary
5000'   1500'   Cretaceous
sea level   Cambrian, Devonian, Mississippian   Pennsylvanian - Permian   Triassic - Jurassic
Proterozoic

10 miles   1 mile   V.E. ~1½

The Great Stack of Rocks in the Grand Staircase (Western Colorado Plateau)

Within a 100-mile transect from Grand Canyon to Bryce Canyon, the Grand Staircase contains a spectacular section of rock over 15,000 feet thick. There is no one place, however, where all of this strata is exposed. A person standing on top of the Grand Staircase at Bryce Canyon has all of the formations present (but not yet exposed) beneath their feet; conversely, a person near the bottom of the Grand Canyon has had all of the layers above previously removed by erosion. The rocks were laid down in numerous environments during nearly 2,000 million years of geologic time. Note that the prominent top cross section is drawn with great vertical exaggeration, whereas the lower section has only 1.5 times exaggeration and more closely resembles the true profile of the Grand Staircase on the modern landscape. The gentle dip of strata to the north, generally less than 2°, explains why the rocks at Bryce Canyon are not 15,000 feet higher in elevation than the rocks at Grand Canyon.

Contents

LIST OF MAPS AND GLOBES

Mesozoic Maps and Globes Continued

Cenozoic Maps and Globes

Future

FOREWORD

Ever since John Wesley Powell nursed his fragile boats through the great canyons of the Colorado River in the years just after the Civil War, geologists have been fascinated by the record of earth history preserved in the strata of the canyon walls, and of the cliffs that rim hundreds of mesas lying back from the river across the Colorado Plateau. The list of pioneer geologists who cut their scientific teeth on the rocks of the plateau reads like a roster of the illustrious American geologists of the late 19th and early 20th centuries. Ron Blakey stands in that tradition, having devoted his professional life to understanding the rock record of the Colorado Plateau.

His research focus should come as no surprise because Ron is a professor at Northern Arizona University (Flagstaff), which boasts the strongest geology faculty at any educational institution located within the confines of the Colorado Plateau. Known the world over for his technical papers on the geologic history of the plateau, he makes his insights accessible to the general reader in this volume of maps showing the shapes of seas and deserts and ancient rivers responsible over geologic time for depositing the strata exposed in the dramatic cliff lines of the Colorado Plateau. The accompanying text by Wayne Ranney, who teaches geology at Coconino Community College (Flagstaff) and is a practiced writer of popular books about geology, sets forth in plain English, unencumbered by geojargon, the backbeat for both the trumpet blasts and the flute music of the maps. Theirs is a happy collaboration.

Each map is a snapshot of an instant or short interval of time during the geologic history of the Colorado Plateau. Geologists call them "paleogeographic" maps—"old" geographies of lost times. What makes them unique in the constellation of paleogeographic maps, which are usually done in the drab pen-and-ink style of the working scientist, is the remarkable computer graphics that have allowed Ron to depict each "old geography" in the same style that the current landscape would be shown in an atlas of the modern world. This makes them vivid and readable by the millions of people who annually visit the scenic wonders of the Colorado Plateau. They will not improve the scenery by so much as one iota, but they allow readers to perceive the deep history that underpins that scenery in ways that only a geologist does. They can also expand the minds of professional geologists themselves, for not even a geoscientist can hold in clear view the kaleidoscopic complexity that the maps reveal as "old geographies" changed through time.

Are the maps fanciful? Yes and no. Yes, because no one trod the landscapes depicted to bring back eyewitness accounts, and artistic license has been used to make each map "come alive" in the mind's eye. No, because each map incorporates the best knowledge that geologists have been able to muster by examining closely the evidence for the nature of the sedimentary strata exposed on the Colorado Plateau. Patient tracing of beds along the cliffs of the plateau has established the lateral extent of each desert dunefield, each paleoriver system, and each incursion of ocean waters that figured in plateau prehistory. It has been a geologic saga spanning hundreds of millions of years to produce the magnificent landscape we see today.

Peruse the maps at your leisure and take them with you on your travels. When you drive the tortured landscape of rock wilderness forming much of the Colorado Plateau, and spot where you are on the maps, you can tell just when that spot of ground was part of a sand desert fully as vast as any in the world today, with dunes stretching beyond the horizon, or when it was part of a shallow seaway with a muddy bottom extending east past where the Rocky Mountains now stand, as far as the Great Plains and up to the Arctic Ocean, or when it was part of an offshore complex of coral reefs and shoals as extensive as the modern Great Barrier Reef off Australia. These and other geologic insights from the maps are at first startling, and always marvelous, but they are as true as the insights that water runs downhill and the sun will rise tomorrow morning.

Knowing the history writ in the rocks can enrich all our lives as we ponder the wonders of the plateau world spread before our eyes, and the maps of this book are an incomparable tool to gain that knowledge. They were produced as a labor of love spread over many years, and use information produced by hundreds, if not thousands, of geologists as the fruits of many decades of painstaking research.

William R. Dickinson
Emeritus Professor of Geosciences
University of Arizona, Tucson
November 2007

Preface by Ron Blakey

Planet Earth is a dynamic body that has been in constant flux for over 4 billion years and continues to reshape itself even today. Distribution and shape of oceans, continents, and seas on the continents; climate, mountains, biota, and atmosphere—all continue to change on this globe that we inhabit. The Earth's long and complex geologic history is recorded in the rocks themselves and in the fossils they contain. This rock record is the main source of data used in compiling this book. Interpreting earth history from such a record is very difficult and awkward to convey to an audience, even for a relatively small part of Earth like the Colorado Plateau. That's where this book comes in.

By combining geologic data in map form with a bit of artistic license, we are able to see a pictorial sequence of ancient landscapes that have existed in the geologic past across the Colorado Plateau region. The maps are prepared so that little explanation, geologic knowledge, or prior experience is needed to read them. Just imagine that you are viewing the Colorado Plateau from the fringes of space and taking snapshots at specific moments in time during a billion and a half years of geologic history.

Detailed paleogeographic maps depicting ancient landscapes have faded somewhat from modern geologic publications. Yet this type of pictorial presentation yields the ultimate synthesis of complex geologic interpretation. Taken as a whole, a long, complex series of landscapes unfolds into a comprehensive geologic history of one of the world's spectacular natural regions.

The methods employed in preparing paleogeographic maps of the Colorado Plateau are both simple and cumbersome. The simplicity involves the transformation of basic geologic data into some kind of a graphic. Geologic publications abound with such material—maps, cross sections, columns, even photographs. But these portrayals of geologic data are usually interpretable only by geologists. Enter the cumbersome aspect of the preparation—turning geologic data and graphics into maps that anyone can read. Here is how it unfolds: I start with a map with boundaries—in this case, the states and counties of the Colorado Plateau and vicinity. I then define the slice of time to be portrayed, and then plot major geologic features interpreted from the data and graphics. This can include locations of rivers, positions of uplands, shoreline trends and locations, positions of dune fields, and so on. Directional data such as paleocurrent flow directions, fining direction of grains in river systems, distribution of mineral grains in sandstone, distribution of floral and faunal elements in the rocks, and patterns of sediment such as sand versus mud all aid in refining paleogeography. Interpretations of rock aspects, such as minerals present and fossil types, are incorporated with data discussed above to make inferences of past climate. This is where the choice of color for the maps comes in; most viewers naturally interpret warm tones as arid settings and cooler tones as suggesting humidity. Finally, with all data portrayed as simple sketches with a few brief words, I am ready to create the ancient landscapes. I doubt that there are many ancient landscapes that lack some counterpart on Earth today (although scale may vary drastically). Therefore I clone modern topography from present digital maps. You might find a piece of the Alps on the Oligocene map or a part of the plains of Argentina on the Moenkopi map. The tributaries to the Ganges and Bramaputra rivers adorn the Triassic and Jurassic landscapes. Because this is done digitally, size, shape, color, and contrast, can all be changed. Because many landscapes are fractal, gorges of East Africa can be manipulated to look like lowland plains of the Cretaceous. However, one goal remains at all times: to portray ancient landscapes within all the boundaries imposed by current geologic data. The final result is that each map shows what a past time might have looked like, not what it actually was—the latter is impossible without H. G. Wells's time machine!

PREFACE BY WAYNE RANNEY

I first met Ron Blakey in January 1979 when I enrolled as a nascent geology student in his Historical Geology class at Northern Arizona University. At the time I had absolutely no idea how profoundly that single class would affect the course of my career. Both Ron and I were very young then but he seemed like a giant to me, well beyond his 6-foot-7-inch physical frame. Ron's lectures concerned the way continents and oceans had constantly reorganized themselves and how Earth's story could be read from sedimentary rocks. In that single semester and with the guiding light of Ron's obvious passion for the history of our planet, I too became captivated with how Earth's landscapes had recreated themselves over and over through geologic time.

At the time of my studies, Ron was in the early stages of describing a "new formation" in the Sedona area. His enthusiasm so infected me that I chose a thesis topic in that area over my beloved Grand Canyon. From that thesis grew the idea for my first book, *Sedona Through Time*. All of Ron's new and emerging ideas about the Schnebly Hill Formation were included in that book, and eventually it was his nomenclature that became the familiar language of Sedona's landscape. Ron's ideas about the Schnebly Hill Formation were formalized at the time I graduated from NAU and, in many ways, I grew up "geologically" with that formation. I felt as though I had witnessed the birth and maturation of a close friend.

Ron moved on to making paleogeographic maps. I went on to become a geologic lecturer on expedition cruises to places like Antarctica and the Amazon. Ron spent several decades perfecting his maps and I became an author. Although apart, we both grew professionally. We met occasionally at professional meetings and I always inquired how his book of maps was progressing. Invariably he'd shake his head and relate to me the last publishing detour. Like many other geology professors, I was saddened to hear of it because I wanted that book for the classes I was teaching at Yavapai College in Prescott, Arizona. At the 2005 annual meeting of the Geological Society of America in Salt Lake City, I asked Ron yet again about his book project. In a response that was becoming all too familiar, Ron shuffled his feet, rolled his eyes, and got a pained looked on his face as he related his last unfortunate defeat. It was at that meeting that I suggested that we collaborate on getting it published.

You are holding the fruit of that fateful conversation and of a thirty-year relationship during which I have been honored to call Ron Blakey mentor and friend. This book is his. I am in awe of his achievement in bringing these wonderful, ancient landscapes vividly to view. We all owe Ron a debt of gratitude for his professionalism and passion, but no one more than I.

Acknowledgments

The authors would like to thank the following people for their help in bringing this book to fruition. William R. Dickinson graciously wrote the book's foreword. Lon Abbott, Terri Cook, Ralph Lee Hopkins, Robert Fillmore, Carl Bowman, Judy Bryan, and Ellen Seeley provided peer review of the final manuscript; David Best and Greer Price reviewed an earlier version. This book would not have been possible without the guidance and leadership of the staff of the Grand Canyon Association; Pam Frazier provided initial editing, and Todd Berger served as editor through the publication phase. We thank them both for their professionalism and desire to see the book succeed. Brad Wallis, the former executive director of the Grand Canyon Association, provided guidance, encouragement, and kept us on track. Ron Short designed the book artistically and with great care to make sure everything was placed just right. Helen Thompson oversaw the marketing and publicity and kept the book moving forward. Ron Blakey would also like to thank his faculty colleagues at Northern Arizona University for encouragement and advice, and especially his wife, Dee, who provided support, encouragement, and patience, even during the times when it appeared that this book might not happen. Wayne Ranney thanks the many professional colleagues he has the pleasure of working with in the field. He also thanks his many students at Yavapai College and Coconino Community College, who through the years have urged him to continue to share his passion and love for the ancient landscapes of the Colorado Plateau.

INTRODUCTION

The brilliantly colored rocks of the Colorado Plateau comprise a truly spellbinding landscape. Within this 130,000 square miles (340,000 sq. km) of rocky bliss there are magnificent escarpments, graceful flat-topped mesas, surreal towers and monuments, and multitudes of deep, sinuous canyons, all framed by an azure sky. Unlike some other notable landscapes on our planet, the Colorado Plateau is dominated by rocks. Not just any rocks, mind you, but colorful, stratified rocks containing subtle textures and minute details that bear witness to the environments in which they originated. These rocks, rich in various shades of red, yellow, orange, brown, tan, purple, and green, are composed of many familiar rock types—limestone, shale, sandstone, conglomerate, mudstone—which manifest as long lines of colorful cliffs or stupendous stair-stepped canyons. When geologists think of paradise on Earth, the Colorado Plateau is what comes to mind.

Rocks dominate this landscape because of an odd and fortuitous combination of geologic factors. First, for hundreds of millions of years this area lay very close to sea level, a place where there was space for sediment to accumulate and be preserved. For much of this time, the region was just above sea level, such that the sediments left behind were brilliantly colored with oxidized iron. Next, the plateau region was uplifted thousands of feet, but in a way that left the horizontal strata mostly intact and flat-lying. All of this was laid bare by the deep dissection of the Colorado River and its tributaries, even though the modern climate is arid. These factors, woven together, explain how this rocky landscape came to be.

Horizontal mesas and plateaus with rectilinear skylines reflect sheetlike sedimentary layers that have been lying here in silent repose since the time of their deposition. Vertical lines originated during the rocks' gentle uplift, and these fractures and cracks allowed flakes of rock to spall off and give shape to the sheer cliff escarpments. Broad, sloped surfaces denote less resistant rock units that have weathered into graceful aprons which are sometimes dissected into spooky badlands. All of these landforms belong very much to the present landscape. However, much of their character belongs to the past. For while the rocks have been shaped recently, the blueprint within those shapes is quite old.

This book is about the blueprint of the rocks. It focuses on events that occurred a long time ago, corollaries found in countless ancient landscapes that represent a record of earth history unrivaled anywhere else on the planet. This history is written in the colorful beds of some extraordinary rocks with interesting names: Tapeats, Coconino, Moenkopi, Chinle. As the story recedes into the past, it tells of the sequential evolution of landscapes that appeared and disappeared through roughly 1,750 million years. The task of comprehending even a small part of this huge story is unquestionably daunting and is one that only a few souls have been privileged to know. While geologists have invented a language which allows them to "time travel" verbally and intellectually with one another in an abbreviated way, this language often leaves the rest of humankind in a hazy cloud of impenetrable jargon. You will encounter some language oddities here; a few are unavoidable. For example, we divide geologic time into eras, periods, and epochs—all with strange but meaningful names.

Above: Admiring the Grand Canyon, one of our planet's greatest exposures of Earth history

Opposite: Then and Now—Triassic Chinle Formation at Gingham Skirt Butte near Paria, Utah. The details preserved in these sediments here and elsewhere allow for the depiction of Triassic geography.

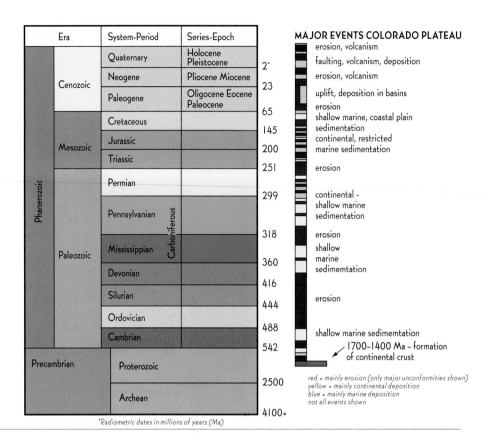

The geologic time scale. Column shows major depositional and erosional events through time.

These time blocks, in turn, are subdivided into lesser blocks such as early, middle, and late. In this book you may notice that these modifiers are sometimes capitalized and sometimes not. Or you might see that they are replaced in some instances with lower, middle, and upper—in which case they are referring to the rocks from those time periods. These details communicate significant nuances to geologists, but they needn't be a stumbling block for the rest of you. We have worked hard to minimize the geojargon.

The paleogeographic maps presented here are not road maps, but rather illustrations that depict what the ancient landscapes of the Colorado Plateau once looked like. They show where ancient oceans once existed in Arizona and where swamps could be found in Utah, and they also suggest the extent of Sahara-like deserts in Colorado and New Mexico. Vast amounts of geologic data have been assembled and synthesized to construct these maps, making previously far-flung, esoteric information readily available to anyone with nothing more than a healthy sense of curiosity. This truly is, we believe, a revolutionary step in making complex geologic information accessible to a wide audience.

We, the authors, have been greatly challenged in this endeavor. As originally conceived, the maps were viewed as being valuable reference tools for professional geologists. Through many stops and starts we became more and more convinced that they were just too attractive and intellectually rewarding for such a narrow audience. For this reason a more inclusive course was charted. Through this process, the maps became more scientifically refined and infused with detail, while the words that explain them were carefully chosen to inspire and educate specialist and nonspecialist alike. Our goal is to provide a window into the past for anyone who is curious about our planet, whether or not they have the rigorous training of a geologist. Our wish is that this book will be a valuable reference work and an avenue for all to the once-privileged information that it contains.

We asked ourselves many times whether it was possible to create a detailed, accurate, and usable reference work for the geologist while making it attractive, interesting, and rewarding to a general reader. We cannot know at this time if we succeeded in our humble attempt to accomplish this end. But we believe that if this seemingly impossible task could ever be realized, these maps could do it. Ever since our species descended from the trees in Africa and looked across those unknown grasslands, we humans have sought out the maps that could show us the way. Here is a book that can show you the way to the history of our planet as recorded in some pretty amazing rocks.

We hope that you will be moved by these maps and the story they tell. We are in awe of this magnificent landscape called the Colorado Plateau.

Both of us have been privileged from very early in our careers to have the plateau as our ultimate mentor and teacher. In many ways this landscape and its story have shaped and guided our lives. We would both be lesser souls were it not for the way a red sandstone cliff simply inspires us to look deeper into its past, seeking out its cryptic message that simply whispers to us, *Things change.* Neither of us had a clue that something as modest as an ordinary rock would stir our passion, yet here we are, giddy at the prospect of sharing what we know. We hope that you also will be moved as you come to understand the deep beauty and earthly pageantry that lies beneath the surface of the Colorado Plateau. Happy time travel!

USING THIS BOOK

The paleogeographic maps found in this book can be a useful tool for the professional geologist, and a fun and interesting way for other readers to sense how earth scientists view the earth through time. To anyone with a modicum of curiosity about the landscape of the Colorado Plateau, this book will truly amaze and inspire. We believe that it reaches across a broad spectrum of interests from skilled specialists to curious travelers who may frequent the many parks and monuments located here. Therefore, we offer some suggestions on how best to navigate your way through the book.

Like any book, this one can be read in the traditional way from front to back. This will take the reader on a chronological journey through time upon the plateau landscape. Used in this manner, the story unfolds with the creation of the oldest rocks and ends with the youngest landscape-forming events. Alternatively, the text has been written such that each chapter stands on its own as a kind of "story with a story"—the chapters make sense without reference to previous text. (This is useful when visiting a particular park or monument with the book in tow by simply referring to the chapter that explains how the rocks found there were formed.) Because of this writing technique, you may find passages that seem redundant, especially when read in the traditional way. We think this is a small price to pay since it provides the reader with more freedom to roam between the various slices of time. If you accept this small amount of redundancy as a given in the text, then you will find it to be a great benefit when casually thumbing through the book in your favorite national park.

No matter how you read the book you will find beautiful photographs, all taken by the authors on their numerous field trips, and interspersed liberally with the text and maps. Each photograph contains a caption that is not numbered: we did not want to interrupt the flow of the text by referencing photographs repeatedly. Each photo and caption explains a concept and is located at the most appropriate place within that chapter. Accompanying the maps and photographs are diagrams, cross sections, and charts that also help explain the geologic ideas.

Almost every rock formation found on the Colorado Plateau has its own map. These maps show how the geography looked when that sediment was accumulating. Some time periods preserve stupendous detail in the rocks and thus have many maps included. A few time periods have little or even no rock record at all, and the geography shown must be approximated and based primarily on what preceded or followed them. The maps can be found in groups of closely related formations within their respective chapters. This makes it easier to see how the landscape changed ever so slightly through time. The global maps add a sense of connectedness between the immediate story of the plateau and the larger events occurring elsewhere on the planet. Remember, the maps can only display a single snapshot of time during the period of rock formation; the real story involves the sequential piecing together of countless snapshots through time.

The book continues with a brief guide to the geologic history of the Grand Canyon and Grand Staircase region. A final chapter shows where you can see the rocks in their glorious splendor. We included twenty well-known and lesser-known areas that we think are the best places to discover the secrets of earth history. An appendix explains how the paleogeographic maps were created. A glossary, list of references, and an index are also included.

Detailed paleogeographic maps depicting ancient landscapes have faded somewhat from modern geologic publications. Yet this type of pictorial presentation yields the ultimate synthesis of complex geologic information. Taken as a whole, a long, complex series of landscapes unfold into a comprehensive geologic history of one of the world's most spectacular natural regions. Understanding the intricacies of a given landscape is not easy, even in the barren rockscapes of the Colorado Plateau. Why is that butte there? Why does the cliff end here or change its aspect of slope? Although these questions relate to geologic history, the landforms of the present landscape are not the topic of concern here; rather this book deals with the origin of the rocks, especially the layers of sedimentary rocks that comprise most of the Colorado Plateau.

A NOTE ON THE DATA
AND INTERPRETATIONS

The geologic events portrayed in this book are based on the most recent and authoritative professional studies on the geology of the Colorado Plateau. Most have been published in peer-reviewed journals. In rare instances, some of the details remain controversial or unresolved, and involve conflicting ideas about the environmental origins of some rock units, or, even rarer, the correlation of some rock units. Obviously, if the origin or relationship of rock units is in doubt, an environmental interpretation is also problematic. Most interpretations, however, are resolved and virtually all geologists will agree with the ones rendered here. Correlation of rock units across the Colorado Plateau is facilitated by the excellent and often continuous exposure of lateral rock units; it is hindered by the lack of fossils in many of the formations. Virtually every rock unit on the plateau has been formally named, most more than one hundred years ago. This formal nomenclature is a worthwhile and useful tool in organizing your view of the plateau's history. We recommend that you become familiar with it, as it neatly arranges the various names for the rock units and is consistently used on the maps, diagrams, and photos.

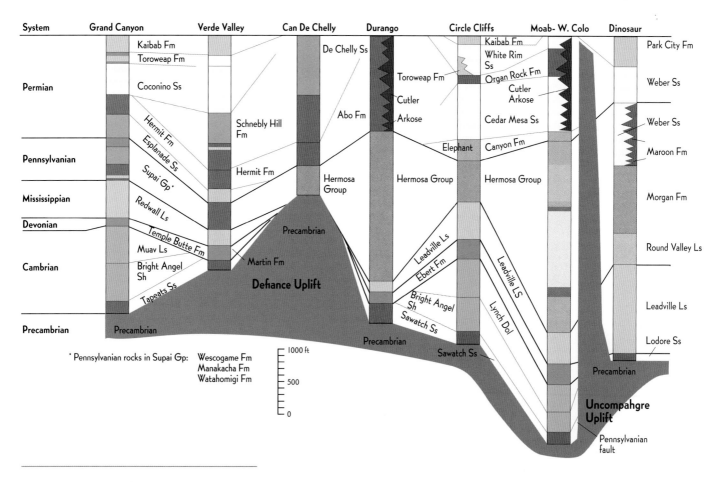

Columnar cross sections of Paleozoic rocks showing the stratigraphic nomenclature and correlations used in this book

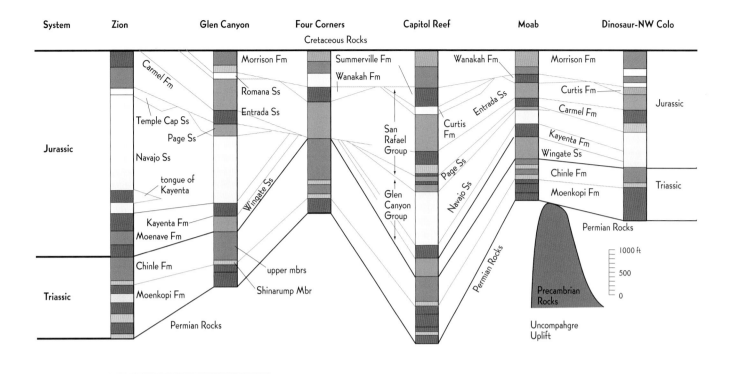

Columnar cross sections of Triassic and Jurassic rocks showing the stratigraphic nomenclature and correlations used in this book

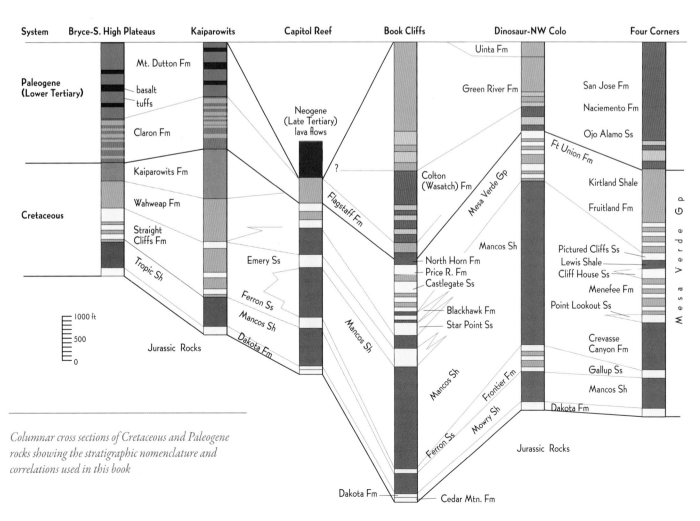

Columnar cross sections of Cretaceous and Paleogene rocks showing the stratigraphic nomenclature and correlations used in this book

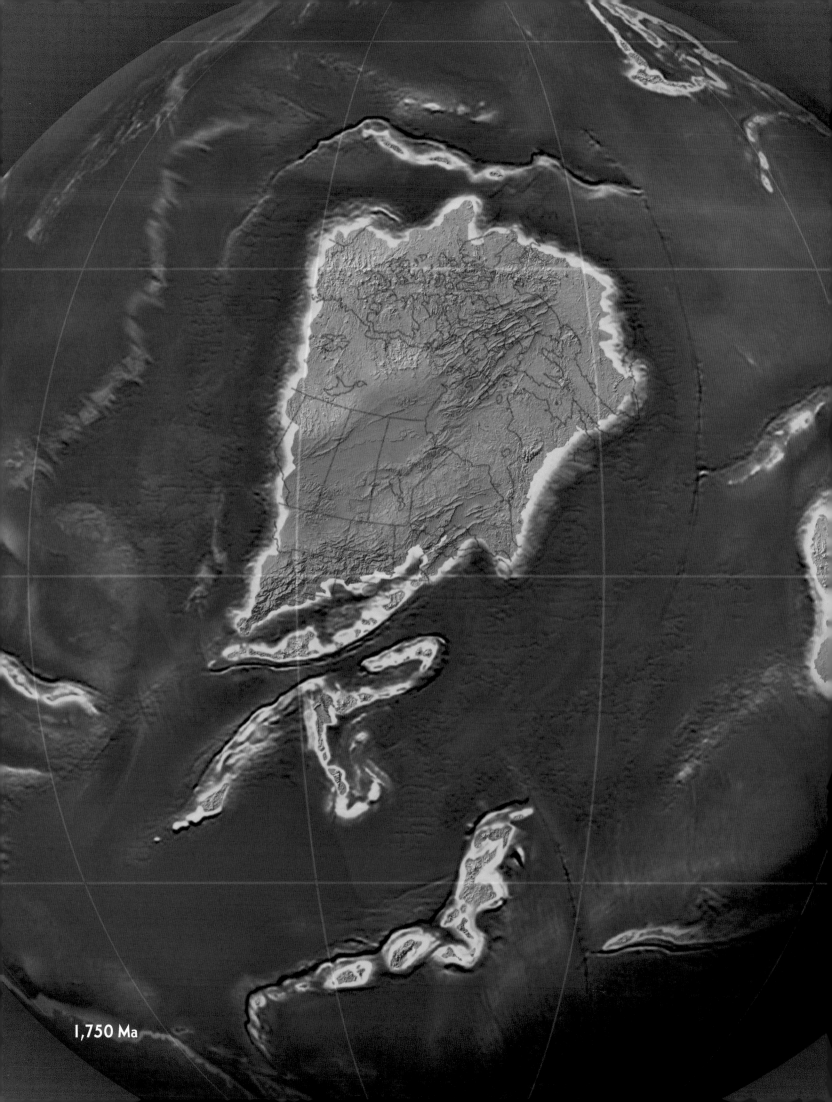

1,750 Ma

FOUNDATIONS OF THE COLORADO PLATEAU
EARLY PROTEROZOIC: 1,750 MILLION YEARS AGO

The long and colorful history of the Colorado Plateau begins with the assembly of its basement or foundation. It may be odd to think of landscapes as having a basement but, geologically speaking, continents cannot exist without one. Oceanic crust, rich in heavy iron and magnesium, is not the kind of material upon which continents can form, even though it is thought to have been what formed the earth's early surface. Through time, our planet has gradually regurgitated the relatively light and frothy material we call continental crust. This crust, rich in silica and aluminum, floats on the earth's hot, roiling interior and acts as the platform upon which sedimentary rocks accumulate. The word "basement" is apt for the plateau's oldest rocks for two reasons: no other rock types can possibly exist beneath them and every stratified layer in the sequence that follows is piled on top of them. Although the basement on the plateau is still mostly buried and concealed beneath younger rocks, it literally supports the plateau's multitude of colorful rock layers. This then, is the stuff of continents—a foundation covered in layers of sedimentary rocks.

It may be incredible to think that prior to 1,750 million years ago (the same as 1.75 billion years), none of the continental crust composing the future Colorado Plateau existed. (Remember that the earth is 4,600 million years old, so 1,750 million years might be considered young in this sense.) Only ocean crust existed in the region and this cannot be considered a part of North America, which, like all continents, was assembled incrementally by the gradual accretion and addition of other pieces of continental crust. Before 1,750 million years ago the southern edge of North America extended only to the region near modern-day Utah and southern Wyoming.

About this time a chain of volcanic islands, similar in distribution and shape to modern Indonesia, began to approach the southern edge of North America. Through a fantastic sequence of geologic events these volcanic islands would eventually compose the continental crust of the Colorado Plateau. As the underlying mantle within the earth slowly churned and roiled, it rafted the volcanic islands northwest toward the old continental edge. Sand and mud was washed from both of these terrains into an intervening ocean basin. At times, volcanic ash or lava erupted and mixed with the sediments, all of which were subsequently buried and lithified. As the island arc slowly drifted closer to the continental edge, the thick stack of buried basin deposits—sandstone, shale, and volcanic rock—became folded, squeezed, and deformed. This pile of rock was then subjected to great heat and pressure at depth.

Opposite: A global view of ancient North America during the Early Proterozoic, about 1,750 million years ago; hereafter, this unit of measurement is abbreviated as Ma [mega annum]. Note the faint outlines of states and Canadian provinces that show what parts of the ancient continent were dry land. The central line represents the equator.

When the volcanic islands eventually collided with North America, the former basin deposits were folded accordion style and placed at great depths within the crust. In this environment the rocks metamorphosed into schist, gneiss, and amphibolite. Certain minerals within these rocks can grow only at specific temperatures and pressures which exist today between ten and fifteen miles (16 to 24 km) within the earth. The presence of these minerals in ancient rocks thus provides the clues to know the depth at which the rocks formed. During this process, the rocks became sutured or attached to the continent and became part of its basement.

At these depths, considered the midcrustal level, rocks were not hot enough to melt but were severely altered while remaining in their solid state. At even deeper crustal levels, the rocks were subjected to much higher temperatures and pressures, and eventually these melted. Huge blobs of liquid magma then rose buoyantly from these depths and forced their way into fractures and foliations in the metamorphic rocks above them. As the magma intruded the schist, it left an artistic suite of dark metamorphic rock pervaded by light-colored igneous rock. Although these rocks presently underlie all of the Colorado Plateau, they are visible only in places where deep dissection of the modern landscape has cut the earth open.

Geologists are able to look at this metamorphism and detect the protoliths (original rock types) that previously existed here. Historically, these rocks were believed to have formed in a Himalayan-type setting by the collision of two large continents. However, detailed studies of the rocks allowed for a new interpretation of the protoliths as originating in an island arc. These rocks were formerly called the Vishnu Schist and Zoroaster Granite, but have a different name today: the Grand Canyon Metamorphic Suite. Within this suite is the Vishnu Schist (altered sandstone and shale), Brahma Schist (metamorphosed basalt), and Rama Schist (altered rhyolite). Understanding these subtleties allows for a more specific interpretation of the environment of deposition. The term "Precambrian crystalline rocks" is generally accepted when referring to the entire group of metamorphic and igneous rocks.

Extremely dynamic events help explain how these midcrustal rocks were eventually brought back to Earth's surface. An orogeny (mountain-building event) that resulted from the arc collision thickened the newly formed continental crust in two dimensions: by growing mountains upward into the sky and by pushing the lower parts of the crust into the soft, pliable mantle. The middle section of this overthickened crust is where rock was altered into schist. Upper-crust rocks found near the mountain top (and thus relatively unaltered), were then subjected to erosion.

Near Phantom Ranch, at the bottom of the Grand Canyon, is the Vishnu Schist. These vertical foliations are the metamorphosed remnants of formerly layered sediments.

Oppostite: The postulated Early Proterozoic paleogeography of the Colorado Plateau region. The map shows an island arc approaching North America in Colorado and the Grand Canyon area. This tectonic setting may have resembled what is found today in Southeast Asia.

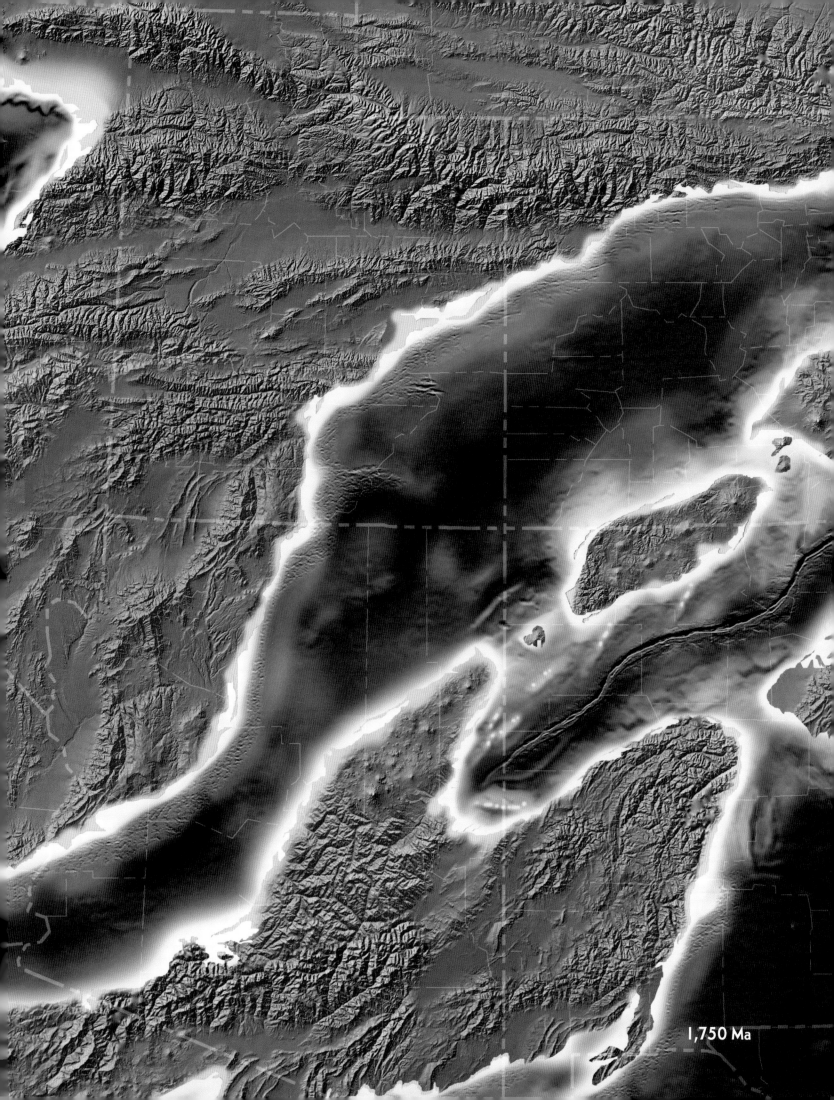
1,750 Ma

Tightly folded Vishnu Schist exposed along the Colorado River in the Grand Canyon. Similar rock types are found in Black Canyon of the Gunnison National Park, Colorado; the Defiance Plateau near Canyon de Chelly National Monument, Arizona; and Westwater Canyon, Utah.

This process eventually brought the midcrustal rocks to the surface in two ways: first by simply removing the overlying rocks, and second, by removing the confining pressure that kept the lower portions of the crust pushed down into the mantle. As erosion continued to remove rocks from the mountain top, the confining pressure was removed as well, such that these midcrustal rocks could rise buoyantly upward, promoting even more rapid growth of the mountain tops, which in turn facilitated even more erosion. This circular dynamism—surface rocks squeezed and folded into the mid levels of the crust, only to return to the surface by mountain building and erosion—wrote the script for the story of the plateau's basement rocks.

Location of Proterozoic Crystalline Rocks

Proterozoic rocks are exposed at the surface in only a few selected localities on the Colorado Plateau: the Grand Canyon and the Defiance Plateau in Arizona; Westwater Canyon in Utah; the Uncompahgre Plateau (including Unaweep, Dominguez, and Escalante canyons), Colorado National Monument, Black Canyon of the Gunnison National Park, and the San Juan Mountains in Colorado; and the Zuni Mountains in New Mexico. Many areas immediately adjacent to the Colorado Plateau expose Proterozoic rocks, including the Rocky Mountains in Utah, Colorado, and New Mexico, and the Basin and Range Province in Utah, New Mexico, Arizona, and Nevada. Some of the specific locations include the Wasatch and Uinta mountains, Utah; the Sangre de Christo and Sandia mountains in New Mexico; the Sierra Ancha and Mazatzal mountains, and the Cottonwood and Grand Wash Cliffs, Arizona; and the Virgin Mountains and Frenchman Mountain in Nevada. Though they are not directly on the Colorado Plateau, these rocks are related in every way and help geologists understand the larger setting of the plateau's Proterozoic history.

Chapter Summary

Although not widely exposed in the modern landscape, the Colorado Plateau is underlain by basement rocks, which formed about 1,750 million years ago. This foundation is composed of both metamorphic rocks (schist, gneiss, and amphibolite) and igneous rocks (quartz dikes, pegmatites, and granite), yielding cryptic but decipherable clues to how, when, and where the rocks were formed. They suggest how ancient sediments and lavas were buried to great depths as volcanic island arcs collided with the North American continent. In the process, new continental crust was formed which was ultimately attached to the continent. All sedimentary sequences that follow are underlain by these rocks.

The Inner Gorge of the Grand Canyon, where the lighter-colored Zoroaster Granite has intruded the slightly older schist

Metamorphic and igneous rocks make up the foundation of the Colorado Plateau. These rocks are rich in crystalline minerals such as quartz (shown here) and feldspar, as well as iron-rich minerals such as mica and hornblende.

1,100 Ma

CHAPTER TWO
ANCIENT SEDIMENTS, UNCONFORMITIES, AND RIFTING
MIDDLE AND LATE PROTEROZOIC: 1,255 TO 740 MILLION YEARS AGO

Evidence suggests that a great thickness of the Precambrian crystalline rocks was eroded between 1,300 and 1,255 million years ago. At that time the plateau region was well above sea level but was gradually being worn down. These 45 million years of erosion eventually planed a flat surface near sea level and set the stage for the deposition of the plateau's oldest surviving sedimentary rocks, the Grand Canyon Supergroup. These rocks are more than 12,000 feet thick (3,700 m), making them more than three times thicker than the significantly younger sedimentary rocks that make up the bulk of the Grand Canyon's walls today! This pattern of hundreds of millions of years of erosion interspersed with hundreds of millions of years of deposition, was repeated often on the plateau.

Supergroup is a term utilized to refer to several time-related rock formations. This practice of lumping many formations together simplifies the terminology and is more commonly used with older packages of rocks. The Grand Canyon Supergroup consists of nine different formations, of which seven combine into two subgroups, called the Unkar and Chuar groups. The Unkar Group, a collective term for five different formations, is composed of, in ascending order, the Bass Limestone, Hakatai Shale, Shinumo Quartzite, Dox Formation, and the Cardenas Lavas. Dikes and sills which intrude the Unkar Group may have erupted the Cardenas Lavas. Next is the Nankoweap Formation, sandwiched between the two groups. On top of these is the Chuar Group composed of the Galeros and Kwagunt formations. On top of it all is the Sixtymile Formation belonging to neither of the two groups. Thinking of all these names, for so many different formations, it's easy to understand why geologists use terms such as group and supergroup.

740 Ma

Opposite: A global view of the Rodinian supercontinent. Its amalgamation may have created a downwarped basin where the Grand Canyon Supergroup rocks accumulated. Its eventual rifting may be what caused these rocks to become faulted, tilted, and mostly eroded.

Right: A global view of the final stages of the breakup of Rodinia. The youngest portion of the Grand Canyon Supergroup was deposited at this time.

The Unkar Group and Nankoweap Formation were deposited between 1,255 and 1,000 million years ago. Evidence suggests that these deposits may have been directly or indirectly associated with a continental collision event in eastern North America called the Grenville Orogeny; they certainly were deposited at the right time for a possible association. This mountain-building event helped to create a supercontinent called Rodinia, of which more will be said later. The Chuar Group is younger than 1,000 million years and the Sixtymile Formation is as young as 740 million years. To keep time in perspective, the span between the oldest of these Supergroup rocks and the crystalline rocks that underlie them is about the same as the time represented from the Cambrian Period (when complex life first evolved) to the present, about 500 million years!

A cross-cutting intrusion, called a dike, cuts through the Proterozoic Hakatai Shale of the Grand Canyon Supergroup. Dikes like this formed in the pluming system of volcanoes that fed the Cardenas Lavas.

The Supergroup and related sedimentary rocks formed mostly in ancient river, coastal plain, and shallow marine settings. Rock types include sandstone, mudstone, limestone, and conglomerate. Fossils of primitive algae are the only evidence of life that has been found in them and the fossil record of similar-aged rocks around the world indicates that life on Earth at this time consisted of simple-celled organisms. The volcanic rocks are of basaltic composition and occur as dikes, which intruded the lower formations, and lava flows which erupted on the surface of a coastal plain during Dox Formation time, about 1,100 million years.

An unconformity, or gap in the rock record, occurs where the Supergroup rocks lie directly on top of the crystalline rocks. This unconformity marks the time when the crystalline rocks, which formed ten to fifteen miles (16 to 24 km) underground, were subsequently uplifted into mountains, then brought back down to sea level by those 45 million years of erosion mentioned earlier. As the crystalline mountains wore down, the landscape was planed into an expanse of gently rolling plains and low hills. It was upon this undulating surface that the Grand Canyon Supergroup accumulated. In nearby areas, related rocks were deposited: the Apache Group in central Arizona; the Pahrump Group in Nevada and eastern California (Death Valley National Park); and the Uinta Group in northeast Utah (Dinosaur National Monument).

Opposite: Approximated paleogeographic setting for the Grand Canyon Supergroup in nearshore basin located between North America and a combined Australia and Antarctica. Sediment that eroded from the continent provided the material for the Supergroup rocks.

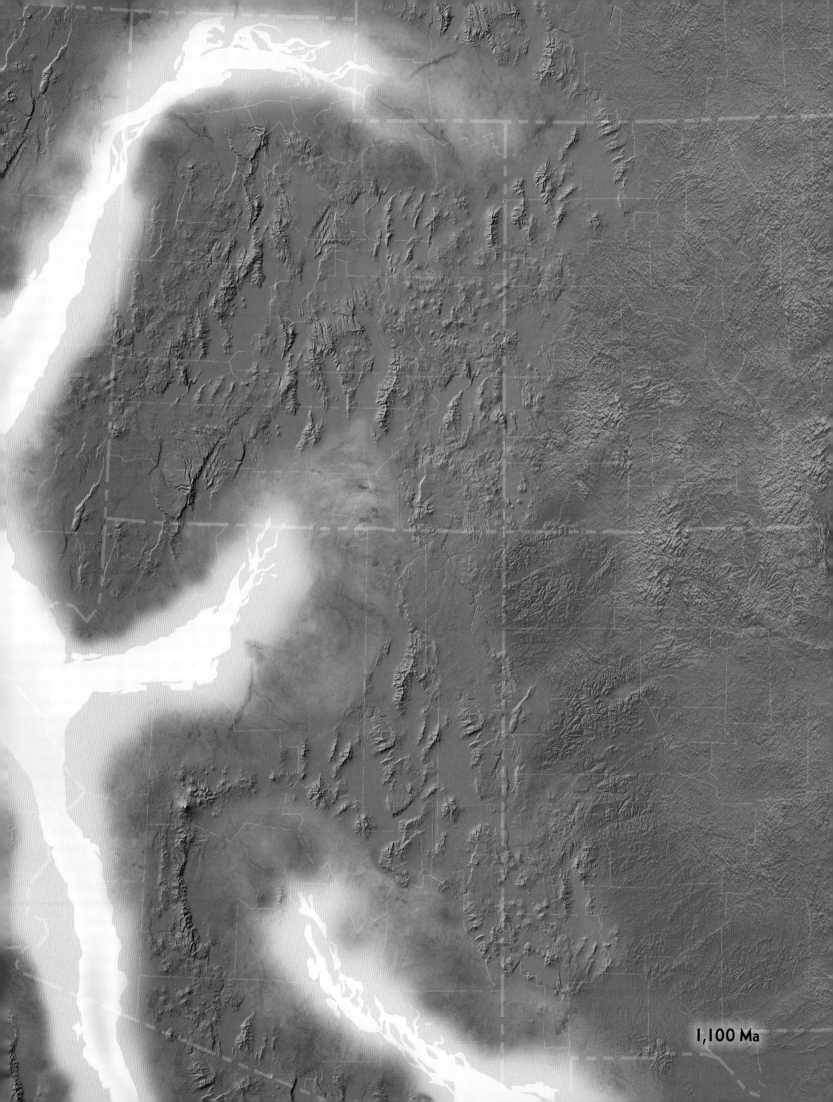

1,100 Ma

Rocks of the Grand Canyon Supergroup, tilted approximately 20 degrees to the northeast (right) below flat-lying Paleozoic rocks, eastern Grand Canyon. This type of break in the rock record is called an angular unconformity.

Because the Grand Canyon Supergroup and related rocks are exposed only in discontinuous outcrops and are curiously tilted, their origins can be difficult to comprehend. Most sedimentary layers on the plateau are generally flat-lying and extend laterally for many dozens or even hundreds of miles. The Grand Canyon Supergroup rocks do not. Their patchwork distribution and tilted character is intimately associated with adjacent faults. These faults disrupted the once continuous and flat-lying strata such that certain blocks were set deeper into the ancient landscape relative to their neighbors. Thus, the few areas that retain remnants of these rocks are found where ancient faulting had positioned the blocks low enough into the landscape that they escaped erosion and were preserved, albeit much faulted and at an angle. Those more expansive parts of the landscape where these rocks are absent represent areas where Supergroup rocks had been uplifted to higher levels, and thus stood in a position where they could be eroded away before the end of the Proterozoic.

The elongate alignment of the remnant fault blocks suggests that a rifting environment, perhaps similar to that found in the rift valleys of East Africa today, is what caused their curious distribution. Rifts form in areas where the earth's crust is being pulled apart or extended by tectonic forces. Western North America was apparently being separated from the large continental mass of Rodinia during Late Proterozoic time. Some evidence suggests that this landmass to the west was a combined Australia, Antarctica, and China, which, together with North America and other pieces of crust to the east, formed the Rodinian supercontinent. Perhaps, then, it was the assembly of Rodinia that was responsible for the deposition of the Grand Canyon Supergroup, and the breakup of that continent that was responsible for their tilted and patchwork distribution. Indeed, the coarse conglomerate found in the uppermost Sixtymile Formation, may record some of the uplift history that marks the end of this exciting time in earth history. This could explain why the Supergroup is preserved only in elongate blocks—rifted valleys?—and why they are gone from most other areas (perhaps the site of uplift adjacent to the rifts).

Angular unconformity in the Grand Canyon

Location of Middle to Late Proterozoic Rocks

Today the hodgepodge remnants of the Grand Canyon Supergroup are spectacularly exposed along the Colorado River and a handful of its tributary side canyons, in the deeper portions of eastern and central Grand Canyon. Equivalent rocks belonging to the Apache Group are exposed in the Salt River Canyon and other nearby mountain ranges in central Arizona, while the Pahrump Group is found in southern Nevada and in Death Valley National Park, California. The Uinta Group is of similar age and is found in the Uinta Mountains at the northern boundary of the Colorado Plateau.

Chapter Summary

The oldest sedimentary rocks on the Colorado Plateau belong to the Grand Canyon Supergroup, a 12,000-foot sequence that includes nine different formations composed of sandstone, siltstone, mudstone, limestone, and basalt lava flows and dikes. These rocks accumulated in nearshore and shallow marine environments that existed in a basin between North America and the combined continental fragments of Antarctica, Australia, and China. This ancient supercontinent is called Rodinia. When this huge landmass began to fragment, the Grand Canyon Supergroup was faulted and tilted, and much of the sequence was completely eroded away. Small blocks escaped destruction, however, and they are found as isolated wedges of colorful strata deep within the Grand Canyon. Significant sections also may exist in the subsurface to the north of the canyon.

Above: Layers of algae once grew in shallow marine environments of the Grand Canyon Supergroup. Here, they accumulated and compacted into fossils forms called stromatolites (shown here).

Below: Geologist pointing to the contact between the Vishnu Schist (below) and the Grand Canyon Supergroup (above) along the North Kaibab Trail, Grand Canyon

FIGURE XX.

11

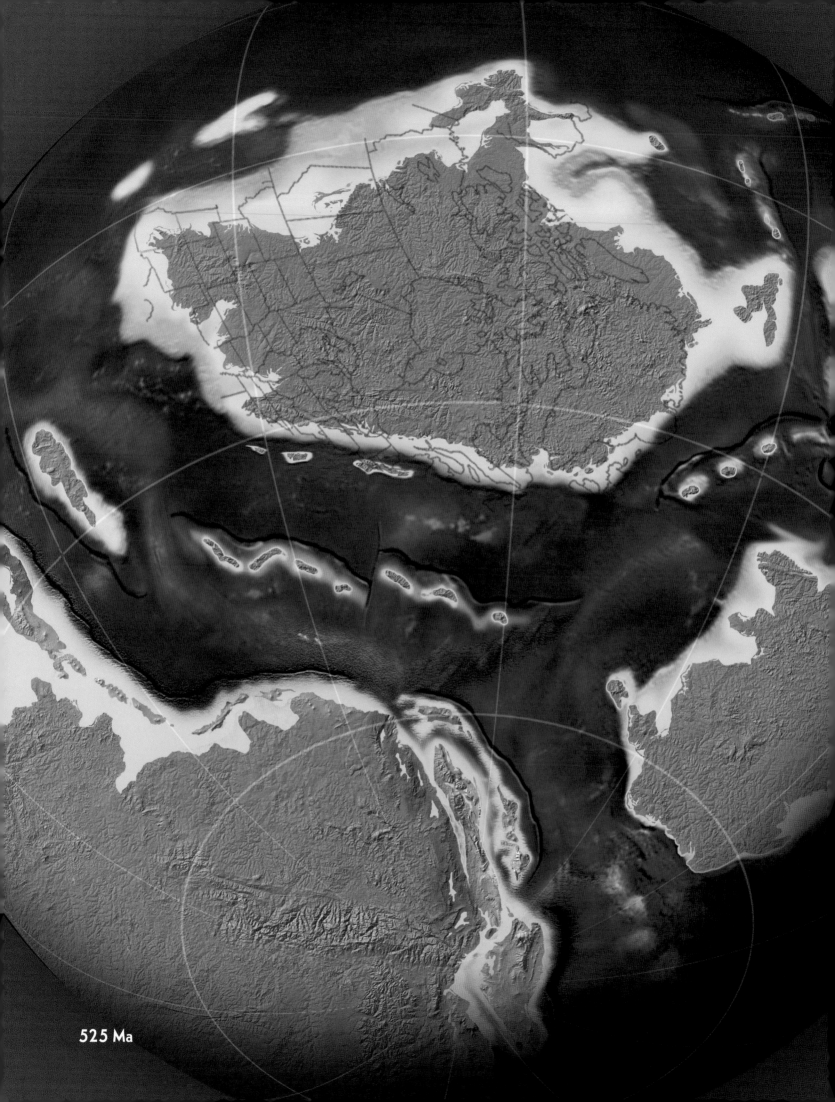

525 Ma

CHAPTER THREE
ANCIENT SHALLOW SEAS
CAMBRIAN TO MISSISSIPPIAN: 525 TO 318 MILLION YEARS AGO

The rifting of the supercontinent Rodinia in Late Proterozoic time split North America away from Australia, Antarctica, and China. This is a phenomenal idea considering how far apart these continents exist today. As the continents separated, the proto–Pacific Ocean was created incrementally and became the site of shallow to deep marine environments. The setting was not unlike the modern-day eastern seaboard of North America. Such tectonic settings are called passive margins because little or no tectonic activity occurs between the continent and the ocean. Passive margins typically undergo rapid, steady subsidence and hence are the sites of great sediment accumulation over geologic time. The Paleozoic Era dawned with passive margin conditions in western North America, which was indeed the site of extensive accumulation of marine sediment. Much of this sedimentation may have been the consequence of the rifting of Rodinia, which would have increased the rate of sea-floor spreading at this time, displacing shallow seawater onto the broad continental shelves.

Such passive margins typically have a hingeline—an elongate partition where subsidence of the crust is much more rapid on the seaward side than the continent side. The Wasatch line, as this Paleozoic hinge is often called in western North America, ran along a tract from southeast Idaho, through Salt Lake City, into southwestern Utah, and through Las Vegas (curiously, much of this line marks the present western boundary of the Colorado Plateau). During most of Paleozoic time, the Colorado Plateau region lay east of the zone of rapid sub-sidence. This resulted in sedimentary rock units of Paleozoic age that are relatively thin on the plateau but which thicken drastically to the west. For example, Cambrian-age deposits are only several hundred feet thick on the Colorado Plateau but expand to about 1,500 feet (460 m) thick along its western border and thicken to nearly 10,000 feet (3,048 m) along the Utah-Nevada line. In fact, the most complete Cambrian section of rocks worldwide occurs in the Basin and Range Province of North America, even though the time period was originally named for exposures in Wales. These relationships document the prior existence of a broad continental shelf (defined as a low-lying margin of continental crust that may be often flooded with seawater), that was crossed many times by repeated transgressions and regressions of shallow seas. This was the tectonic setting for the deposition of early and middle Paleozoic rocks across the Colorado Plateau.

Opposite: The global positions of the continents during the Cambrian Period (about 515 million years ago; hereafter, this unit of measurement is abbreviated as Ma [mega annum]). North America was isolated and straddling the equator at this time and was surrounded by passive margins. Note the position of the Four Corner states in the northwest section of North America.

The Great Cambrian Transgression

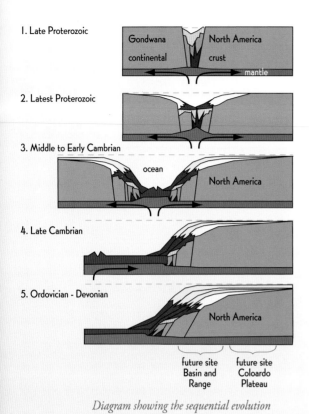

1. Late Proterozoic

Gondwana North America

continental crust

mantle

2. Latest Proterozoic

3. Middle to Early Cambrian

ocean

North America

4. Late Cambrian

5. Ordovician - Devonian

North America

future site Basin and Range

future site Coloardo Plateau

Diagram showing the sequential evolution of passive margin settings in western North America

A resistant knob of the crystalline basement rocks that projected above the Cambrian landscape (note the paleotopography that separates the two sections of brown layered sandstone seen in the sunlit wall of the Grand Canyon's Inner Gorge). This knob was eventually buried in Cambrian sediment, and the contact between these two sequences is known today as the Great Unconformity.

Cambrian rocks were deposited across the Colorado Plateau from west to east as shallow seas of the proto–Pacific Ocean transgressed (flooded) the low-lying continent. In the Grand Canyon region the surface upon which these seas transgressed was variably underlain by remnants of Precambrian crystalline rocks and isolated blocks of the Grand Canyon Supergroup. This surface was generally flat; however, hills of both rock types projected above the mostly planar surface. Linear outcrops of the Supergroup were positioned as elongate blocks of tilted sedimentary units, often capped by resistant remnants of the Shinumo Quartzite. Some of these projections stood as high as 700 feet (213 m) above the surrounding plain before transgression. During transgression, these blocks stood as isolated islands that were ultimately buried in sediment as sea levels rose. A great angular unconformity is preserved where flat-lying Cambrian strata overlie and cover the tilted remnants of the Supergroup rocks.

Elsewhere across this surface, occasional rounded knobs of resistant Precambrian crystalline rock projected a few hundred feet above the coastal plain. These resistant islands of harder rock shed sediment onto the plains below which consisted of beveled schist and granite—the weathered roots of the once formidable Precambrian mountains. It was in this setting that Cambrian sandstone, 525 million years old, came to rest on the crystalline rocks, 1,750 million years old. This contact created the Great Unconformity, a huge gap in the rock record spanning some 1,225 million years of time.

North America had been weathering for the previous few hundred million years, and an abundant source of sediment was readily available as the seas moved eastward. The shallow seas reworked and redistributed the sediment into myriad shallow marine and shoreline deposits. Sand accumulated near the shoreline in high-energy environments (where the currents moved vigorously), such as the Tapeats Sandstone in the Grand Canyon and the coeval Lodore Sandstone in Dinosaur National Monument. Alternatively, mud accumulated in offshore, low-energy environments such as the Bright Angel Shale. Trilobites flourished on these sandy and muddy bottoms, scavenging for food in the shoreline currents. Farther offshore, shallow, sun-drenched sea bottoms were home to many lime-secreting organisms such as brachiopods and mollusks. They flourished in these calm, sediment-starved waters and caused the widespread deposition of Muav Limestone. This specific sequence of sedimentation—sandstone, shale, and limestone, piled one on top of the other—

Opposite: Early to Middle Cambrian setting of the Colorado Plateau during deposition of the Tapeats Sandstone. The sea rhythmically transgressed (advanced) eastward until the Late Cambrian, when much of the land at the eastern margin of the map was covered by shallow seas.

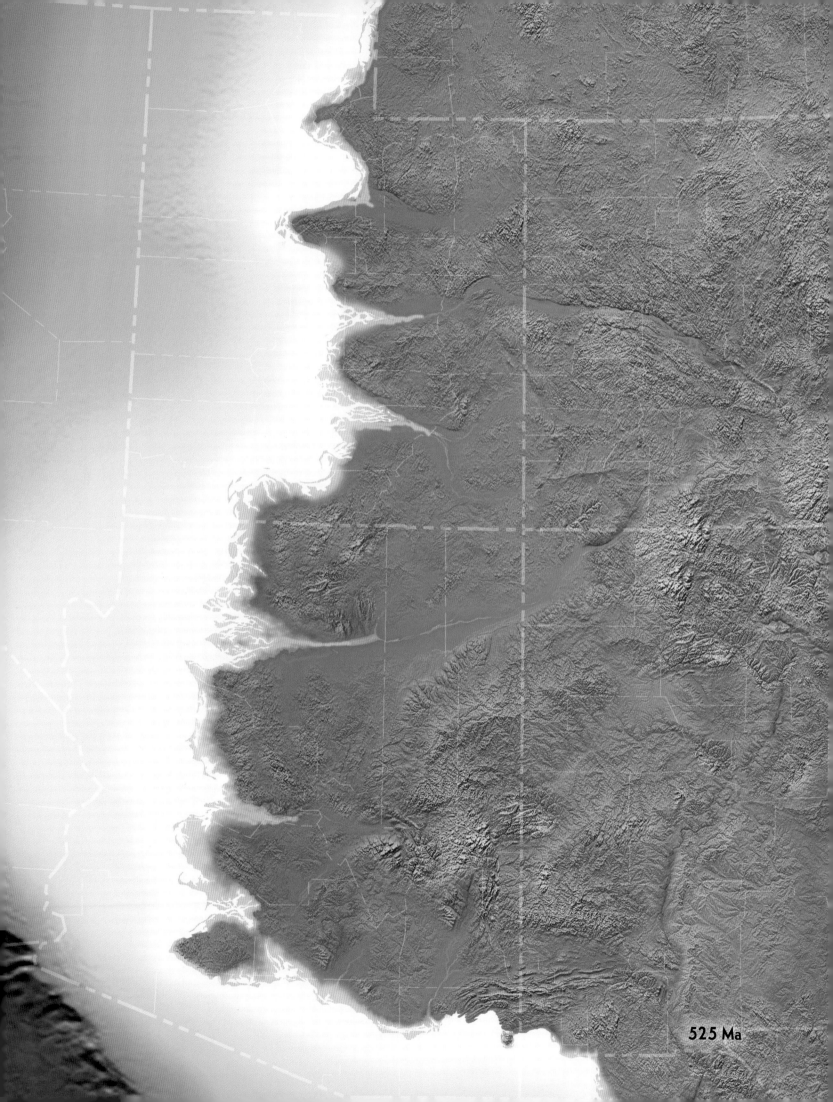

525 Ma

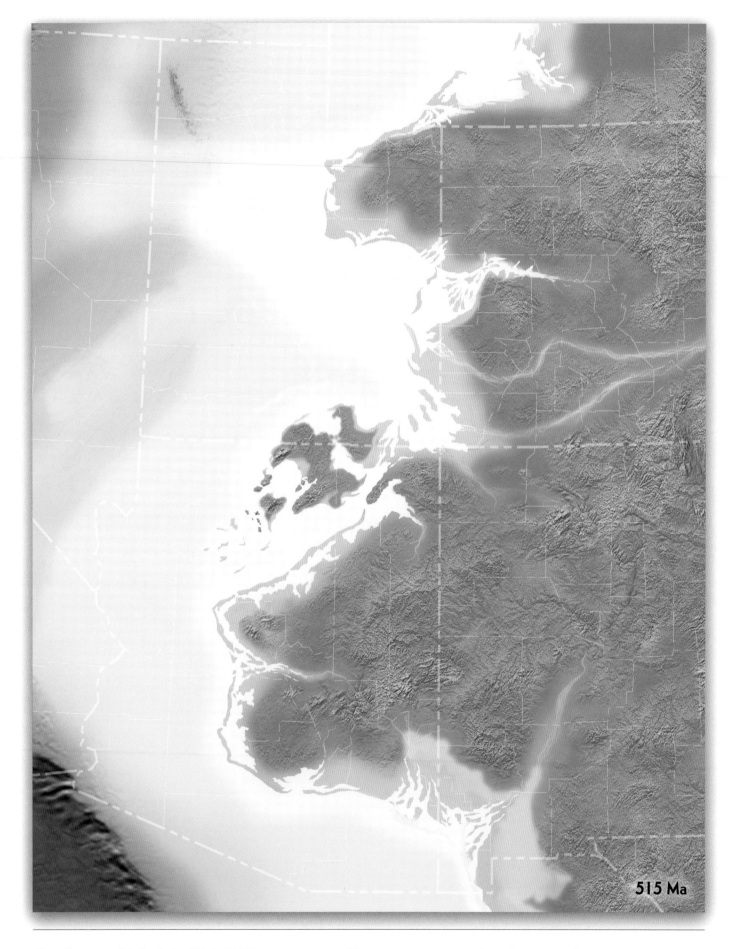

515 Ma

Above: Deposition of the Bright Angel Shale (515 Ma) in more protected offshore environments that were below the storm wave base

Opposite: Conditions during the deposition of the Muav Limestone (505 Ma), which accumulated offshore in areas removed from sand and mud input

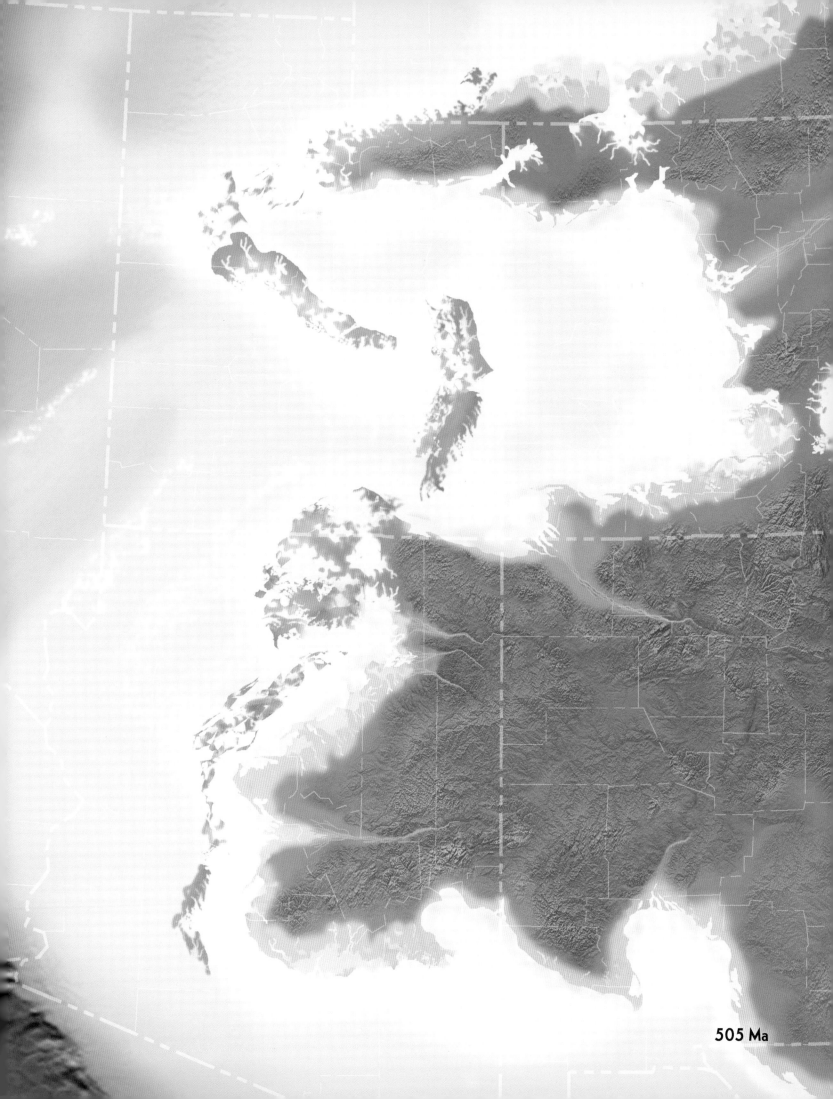

505 Ma

is a textbook example of a marine transgressive sequence formed on a passive margin. The Cambrian rocks of the Colorado Plateau arc often used as the classic example of this type of sedimentation, since their preservation and exposure are excellent. Some curious and subtle deviations from this simple model are preserved in these rocks as well. After a transgression to the east, the sea would occasionally regress to the west a short distance, then transgress again to the east—two steps in, one step out, two steps in, again and again. This sequence produced the complex interfingering or intertonguing of sediment types so well exposed in the Cambrian rocks of the Grand Canyon. Indeed, because of these fluctuating shoreline conditions, it is difficult to draw definitive boundaries between Tapeats Sandstone, Bright Angel Shale, and Muav Limestone. Through time, as the sea relentlessly moved eastward (noting that it may have taken up to 30 million years to transgress the overall width of Arizona), sandstone deposits were buried by mudstone and mudstone deposits were buried by limestone.

Another interesting aspect of this transgressive sequence is that it suggests that a single formation or rock type may be of different ages depending on where it was deposited. For example, Tapeats Sandstone was at one time being deposited in beach and nearshore environments near Las Vegas, Nevada. But as the sea transgressed eastward through time, the Tapeats sediment type moved in step with the shoreline such that it was deposited in western Grand Canyon at a later time and in eastern Grand Canyon at a much later time. The Tapeats looks the same from one place to the other but is of a considerably different age in each location. Fossil evidence bears this out. This means that the plateau's Cambrian-age rock units are time-transgressive; that is, the age of each formation gets progressively younger toward the east.

Lastly, sediment can only accumulate and be preserved if space is available for it to reside. This accommodation space, as it is called, is created if the water deepens through time in a transgression or if the seafloor subsides relative to sea level.

Left: Muav Limestone in the Grand Canyon

Geologic evidence points toward both processes being active during the Cambrian on the Colorado Plateau. However, it is the second process that was far more important than the first in preserving these sediments—passive margins are typically the sites of significant subsidence. This subsidence may result from tectonic forces actively depressing the crust, or it may be the overlying weight of the accumulating sediment that may lower the crust to create the accommodation space. Another possibility in this particular instance is that the earth's crust in this location was cooling after all of the tectonic excitement of the Late Proterozoic Rodinian split.

The Cambrian Explosion of Life

The Cambrian Period witnessed an explosion in life-forms on planet Earth. Whereas most of the Precambrian Era contains only simple-celled fossil algae and bacteria, the Cambrian fossil fauna is replete with multicellular organisms that had intricate shells and advanced, complex organs such as eyes and body appendages. (The end of the Proterozoic Eon, known as the Ediacaran Period, does record the beginning of this greater complexity in life-forms that began about 625 million years ago.) Within 100 million years after the beginning of the Cambrian, all zoologic phyla known today appeared in the fossil record. Again, placing this in perspective, it had taken perhaps 3,000 million years of earth history to produce the first multicellular animals, then only 100 million years to evolve the most complex phyla present on Earth today.

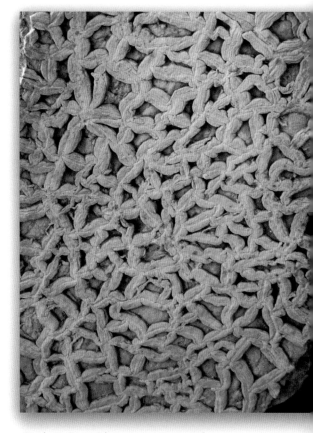

Cambrian "worm" burrows crisscross a surface exposed on the Muav Limestone, Havasu Canyon, Grand Canyon

As one looks at the maps and ponders this Cambrian landscape, it is interesting to keep in mind that all animals during this time were exclusively marine forms, that is, ocean-dwelling—no land animals, or plants for that matter, had yet evolved on planet Earth. As rivers brought sediment to the edge of ancient North America, the setting was one of bare rock landscapes and gravelly or dusty sediment-filled valleys, all without a hint of terrestrial plant or animal life. Landscapes on Mars or in the ice-free valleys of Antarctica may be the best modern analogies to the Cambrian terrestrial landscape of the Colorado Plateau, which certainly would have been a very foreign thing to behold.

Missing Time Periods: Ordovician and Silurian, 488 to 416 Million Years Ago

The next two time periods, the Ordovician and Silurian, have no rock record on the Colorado Plateau. Whether rocks of this age were deposited and then removed by subsequent erosion or never deposited at all cannot be determined. A well-known and widespread period of erosion at the beginning of the Devonian suggests that rocks of this age might once have covered the plateau, only to have eroded away. Most of the Ordovician and Silurian deposits found adjacent to the plateau are carbonates that formed in warm, shallow, limy seas. This implies a continuation of the Cambrian passive margin environments. Almost certainly these Ordovician and Silurian seas lapped at the edges and perhaps even transgressed and covered parts of the Colorado Plateau. (Utah's Beaver Dam Mountains do contain Ordovician-age rocks and these are just a stone's throw from the plateau edge.) Ordovician and Silurian rocks, which comprise about 72 million years of earth history, are common in the midwestern, eastern, and northern parts of North America.

Brachiopods, nautiloids, and crinoids dominated clear, limy Paleozoic seafloors.

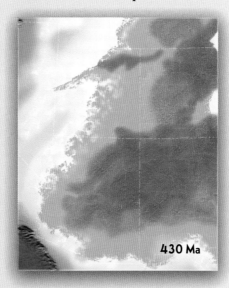

470 Ma   430 Ma   400 Ma

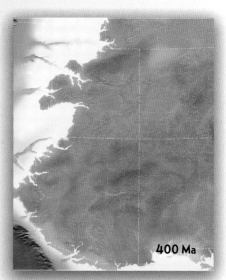

Above left: The Colorado Plateau area in Ordovician time, about 470 Ma

Above middle: The Colorado Plateau area in Silurian time, about 430 Ma. Although Ordivician and Silurian rocks are absent from the plateau region today, they may have once been deposited but eroded away before other layers were laid down.

Above right: Early Devonian paleogeography (400 Ma). Much of the plateau region was high relative to sea level and rocks were either not deposited or were eroded during this time.

Although no deposits remain on the plateau, the Ordovician and Silurian periods were nonetheless momentous times worldwide. It was sometime during these two periods, between about 480 and 450 million years ago, that plants and animals gained a terrestrial foothold. Some of the earliest evidence of this comes from Great Britain and Greenland. North America was apparently home to these early land-dwelling forms but also a quite subdued continent with few mountains and widespread shallow seas ringing its borders. Late in this time period, it collided with pieces of crust now belonging to the Baltic region in modern Scandinavia. This occurred during the initial assemblage of the giant supercontinent of Pangaea, of which much more will be said later.

Carbonate Seas of the Devonian and Mississippian, 416 to 318 Million Years Ago

Devonian rocks are present across much of the plateau region but are often buried by younger strata and are thus known mostly from well-drilling information. Limestone and dolomite are the primary rock types; they document the former presence of widespread clear, shallow seas. Sandstone and mudstone are present as well, denoting those areas where sand and mud were being deposited in coastal settings.

Modern environmental settings can help us understand the relationship between limestone and dolomite (carbonate) deposition, and sand and mud (clastic) deposition. Areas such as eastern Florida, Yucatan, and the Persian Gulf are areas dominated by carbonate deposits on shallow marine shelves. In these areas, where the climate is arid to semiarid and rivers are essentially nonexistent, there is a lack of clastic input. This favors carbonate deposition since very little sand or muddy material is delivered to the shoreline. In contrast, areas with more humid climates or having large muddy rivers, like the modern Mississippi Gulf Coast, deliver large quantities of sand and mud to the shoreline, and there is very little carbonate sedimentation. Therefore, observing where carbonate and clastic depositions occur in today's environment provides clues as to the environmental conditions that may have existed adjacent to these ancient seas.

Carbonates can also yield information about the ancient depth of the water. Many organisms preserved as fossils in carbonate rocks required shallow water in which to live. This is because sunlight is required for photosynthesizing food for marine organisms. Most photosynthesis in modern oceans occurs in the first 100 to 200 feet (30–60 m) of the water column. The fact that many Paleozoic organisms lived on or attached to the bottom suggests that the water depths could have been no deeper. Devonian fossil faunas are diverse, with all major phyla including vertebrates (fishes) represented. Brachiopods and corals dominated the near-equatorial Devonian seas of the Colorado Plateau region.

In the Grand Canyon only one formation, the Temple Butte Limestone, is of Devonian age. It occurs in the eastern part of the canyon as discontinuous deposits set within channels cut into the underlying Muav Limestone. Toward the west, these channel deposits coalesce into a continuous rock layer which is more than 400 feet (120 m) thick at the Grand Wash Cliffs; Devonian deposits thicken even more to the west in the Basin and Range. This pattern again reflects the continental shelf conditions that were in existence along the Wasatch line during the whole of Early and Middle Paleozoic time. The specific environment interpreted for the Temple

Above, top: Devonian deposits in eastern Grand Canyon are only preserved in channels cut into the underlying Muav Limestone. The deposit pictured is located along the Colorado River.

Above, bottom: The Temple Butte Limestone in western Grand Canyon is a deposit more than 400 feet (120 m) thick that records how the continental shelf was subsiding to the west in Devonian time.

Butte Limestone is that of a system of estuary channels in eastern Grand Canyon that coursed to a shallow sea to the west and whose floor subsided increasingly through time. Exposures within the Grand Canyon reveal this ancient configuration of estuaries to the east and ocean to the west.

To the south of Grand Canyon, Devonian deposits are called the Martin Formation; they are exposed near Jerome, along the Mogollon Rim, and as far south as southern Arizona. Elsewhere on the plateau, deep wells have documented the presence of Devonian strata, which are known as the Elbert Formation (a clastic deposit) and Ouray Limestone (shallow-water carbonate) in the subsurface of Utah and in the Rocky Mountains of Colorado.

The Mississippian on the Colorado Plateau was again a time of extensive carbonate deposition. (In other parts of the world, the Mississippian is called the Lower Carboniferous, reflecting a worldwide abundance of carbon-rich coal and oil.) But Mississippian rocks on the Colorado Plateau and much of the Western Interior are vast expanses of nearly pure limestone from Kansas to Nevada and from Canada to southern Arizona. This implies arid conditions for the plateau region, even though it was a time when shallow seas ruled the landscape.

The deposits from this period include the imposing cliffs of Redwall Limestone deep in the Grand Canyon, the Escabrosa Limestone in southern Arizona (Saguaro National Park), the Monte Cristo Limestone in Nevada (Lake Mead National Recreation Area), the Leadville Limestone in the San Juan Mountains and Glenwood Springs area in Colorado, and the Madison Limestone in the northern and central Rockies (Grand Teton National Park and Dinosaur National Monument). All of these formations are coeval and result from deposition in a widespread sea. (Their different names survive because early studies could not know their intimate relationship.) The shallow, clear, limy seas that deposited these rocks received almost no input of sand or mud and thus were favorable sites for the growth of lime-secreting marine organisms. Brachiopods, bryozoans, corals, and echinoderms dominated. Paleogeographic maps for this time show virtually no land in western North America. The continent existed but was submerged beneath these tropical seas.

385 Ma

Devonian globe (about 385 Ma). North America is still equatorial but has rotated counterclockwise. Various pieces of crust have collided with eastern North America to form parts of the Appalachian Mountains. Western North America still has a passive margin, but an island arc approaches from the west (Antler Orogeny, Nevada).

Left: Colonial corals such as this Syringopora are typical fossils found in Devonian and Mississippian rocks on the Colorado Plateau. This specimen is from the Redwall Limestone, Grand Canyon.

Opposite: Late Devonian paleogeography (370 Ma) shows the shallow marine conditions that existed in the Colorado Plateau region. Equatorial seas abounded with calcite-secreting organisms, especially algae, brachiopods, and corals, found in the rock units. Note the submarine canyons carved on the continental shelf in eastern Nevada.

370 Ma

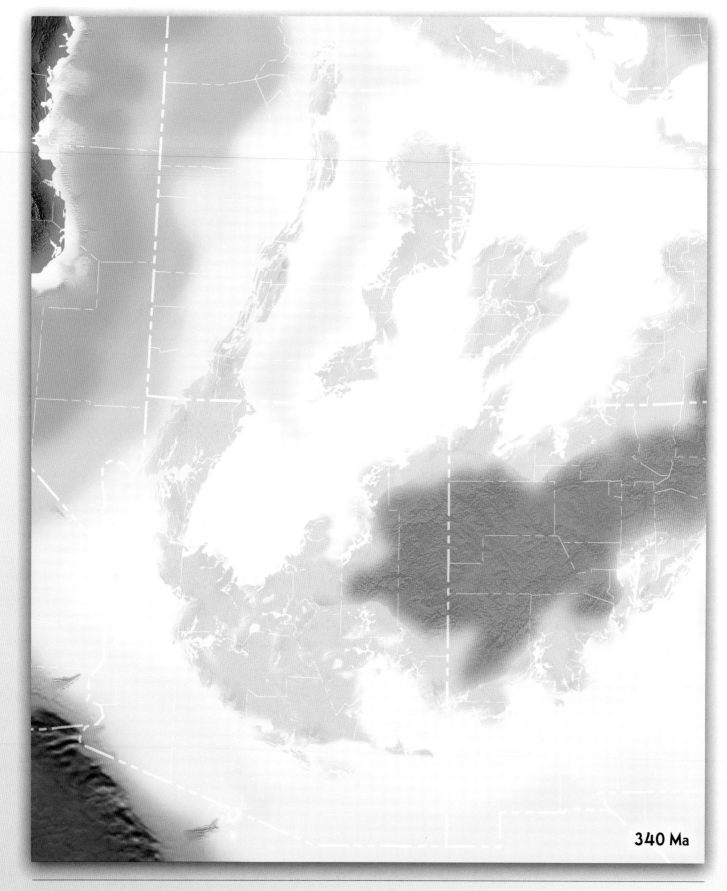

340 Ma

Above: The Mississippian Period (340 Ma) witnessed a dominance of shallow marine carbonate deposition across the region exemplified by the imposing Redwall and Leadville limestones. Note the highlands in Nevada formed as a result of the Antler Orogeny, when an island arc collided and was attached to western North America.

Opposite: During the Late Mississippian Period (325 Ma), the region was uplifted and streams carved valleys into the Redwall Limestone. Several marine transgressions flooded these valleys and left estuary deposits known as the Surprise Canyon Formation in the Grand Canyon.

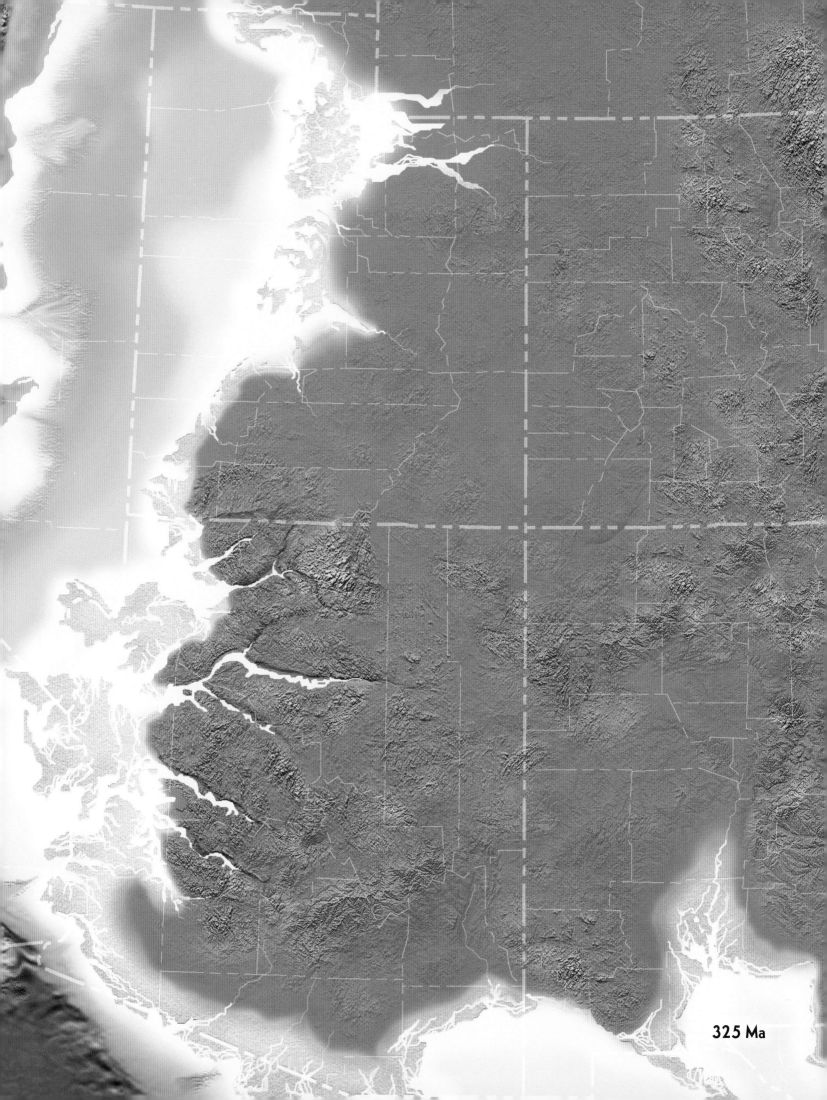

325 Ma

Following the extensive Mississippian seas, much of North America was gently uplifted and the seas withdrew from the interior, causing an unconformity to develop in the rock record. In western Grand Canyon, terrestrial streams carved valleys into the exposed upper surface of Redwall Limestone. Several times during latest Mississippian time, seas flooded these river valleys and left marine deposits interbedded with fluvial sand and gravel. This estuarine setting deposited the Surprise Canyon Formation, which escaped detection by geologists until the 1980s. It remains the newest deposit to be recognized in the Grand Canyon. In other areas of the plateau, the Molas Formation exposes soils that developed on top of the exposed limestone. And to the far west in Nevada, the Antler Mountains were separated by a seaway from the Colorado Plateau region. This type of marine basin is known as a foreland basin. It formed under the weight of the thickened crust in the Antler Mountains, which exerted downward pressure on the adjacent crust, allowing the sea to flood that area.

Top: Aldabra Lagoon in the Seychelles in the Indian Ocean exemplifies the clear, shallow marine conditions in the Mississippian of the Colorado Plateau.

Above: Mississippian-age shark tooth found in Oklahoma

Left: A lens of the Surprise Canyon Formation cut into the Redwall Limestone at Fern Glen Canyon in the Grand Canyon. These channel deposits originated in estuary environments.

26

Location of Cambrian, Devonian, and Mississippian Rocks

Cambrian, Devonian, and Mississippian rocks are exposed in the Grand Canyon and along the Mogollon Rim from east-central to northwest Arizona. They remain in the subsurface throughout the central portion of the Colorado Plateau, but reemerge at the foot of the Uinta Mountains in Dinosaur National Monument and also in the Glenwood Springs–Rifle area of central Colorado. They typically can be found in the same areas where Cambrian rocks are exposed.

Chapter Summary

The ongoing fragmentation of Rodinia at the start of the Paleozoic Era created passive margin environments in western North America. These tectonically quiet settings allowed for the preservation of a classic transgressive sequence upon the slowly subsiding continental shelf. Cambrian rocks become progressively more carbonate-rich through time as the passive margin evolved and seas pushed farther east. Rocks of Ordovician and Silurian age are missing on the plateau, creating an unconformity in the rock record of perhaps 135 million years duration. Whether sediments were never deposited or were initially deposited and removed by subsequent erosion is unknown. Carbonate deposition, however, did return to the area during the Devonian and Mississippian periods. Toward the end of the Middle Paleozoic, the plateau region became completely submerged beneath the shallow water of the Redwall Sea.

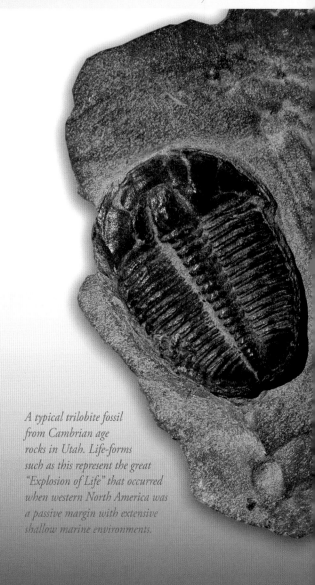

A typical trilobite fossil from Cambrian age rocks in Utah. Life-forms such as this represent the great "Explosion of Life" that occurred when western North America was a passive margin with extensive shallow marine environments.

The Redwall Limestone within the Grand Canyon was formed in shallow marine environments like the photograph opposite top.

300 Ma

CHAPTER FOUR
CONTINENTAL UNREST AND GEOLOGIC CYCLES
PENNSYLVANIAN AND PERMIAN: 318 TO 251 MILLION YEARS AGO

Worldwide, the late Paleozoic witnessed a time of tremendous geologic change on planet Earth. Great continental collisions and crustal unrest assembled a giant supercontinent called Pangaea. The landscape of Pangaea was dominated by arid and semiarid environments, especially along its western edge where the future Colorado Plateau would form. Sedimentary sequences of the Pennsylvanian and Permian time periods differ significantly from the mostly carbonate rocks of the early and middle Paleozoic. Newly uplifted mountains supplied abundant gravel, sand, and mud to the sedimentary environments of the plateau region. This setting preserved sediments from the first of many great fossil deserts on the plateau.

The tectonic setting of the late Paleozoic contrasted as well with that of the early and middle Paleozoic. It was during this time that western North America began a gradual transition from more than 200 million years of passive margin deposition to an even longer period of deposition on an active margin. Indeed, active margin environments continue to this day in western North America. Active margins are those in which the edge of the continent is a plate boundary, in this case a convergent plate boundary marked by the collision of two continents, namely South America and North America, with the crust of the Pacific Ocean. These complex tectonic forces are what fractured and deformed the West's continental crust, and these compressive forces produced one of the West's most striking and important ancient landforms, the Ancestral Rocky Mountains.

280 Ma

The Ancestral Rockies, which began their rise about 315 million years ago and were not completely eroded away until 150 million years later, were in no way related to the modern Rockies except for their location (which explains the name given to both by geologists). They extended from east-central Utah through southwestern Colorado and into northern New Mexico.

Opposite: Pennsylvanian global view of North America shows a part of Laurasia colliding with Gondwana to form the supercontinent of Pangaea. The final phase of Appalachian mountain building is the result. Note that western North America begins to evolve into an active margin at this time, causing portions of the western crust to warp up into mountains.

Above: Permian global view of southern Pangaea. Note the extensive glacial areas in the south polar region.

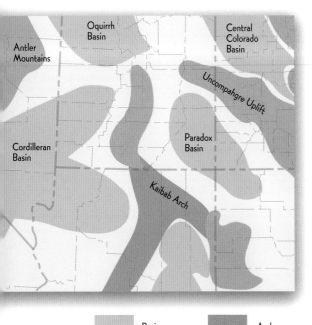

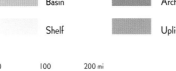

| Basin | Arch |
| Shelf | Uplift |

```
0        100        200 mi
0    100    200    300 km
```

Obviously, no one ever saw these mountains, yet abundant evidence for their existence comes from the voluminous sediments that were shed from them and remain preserved as part of the brilliantly colored stack of sedimentary rocks on the Colorado Plateau. One element of this great system of mountains, the Uncompahgre uplift, particularly influenced deposition on the Colorado Plateau. This massive, uplifted block was stripped of its thin veneer of Paleozoic sediments exposing a core of Precambrian crystalline rocks, the very ones that form North America's basement complex.

When the earth's crust is fractured and uplifted during mountain building, adjacent areas commonly warp, buckle, and subside. These broad areas of subsidence—usually semielliptical in shape—create sedimentary basins, and a number of these basins formed in places adjacent to the Ancestral Rockies. Sediments that accumulated in these basins are generally many times thicker than what is preserved on nearby shelf areas. This is because subsidence in the basin provided the extra accommodation space necessary for the sediments to be "put in storage" as sedimentary rocks; the nearby uplifts supplied the material to fill such basins. Each of the late Paleozoic basins on the plateau had a slightly different shape, source area, and subsidence history, which left a unique sedimentary record. Therefore, each of these basins has its own set of deposits and formation names, while intervening shelf areas between the basins may have no sedimentary rocks at all.

Yet another complexity exerted its influence on the deposition of Pennsylvanian and Permian rocks on the plateau: sea level fluctuated abruptly and repeatedly throughout this time. This occurred in response to a well-known glaciation centered in the southern hemisphere, where southern portions of the Pangaean supercontinent extended over Earth's South Pole. When glaciers expanded, they locked up huge amounts of freshwater on land, and sea levels dropped. As the glaciers melted, fresh water returned to the ocean and sea level rose. Sea level changes of several hundred feet can occur when this happens, an obvious concern with respect to our modern episode of global warming. As many as sixty glacial cycles during the late Paleozoic have been identified within the rocks. Indeed it is a combination of this sea level, uplift, and sedimentation history that explains the great degree of preservation in the rocks for this time period, and thus the sheer number of maps made possible by these histories. (Note that the remaining maps in this book are spaced in time at an average of only 12 million years, with many separated by just 5 million years.)

Left: The Pennsylvanian rocks of the Colorado Plateau slowly changed from marine rock types to continental rock types. This view is from the San Juan River canyon in southeast Utah, and the transition interval is in the stair-stepped lower canyon.

Shallow Seas, Coastal Plains, and Rivers: Pennsylvanian

The Pennsylvanian time period, officially a subperiod of the Carboniferous, is divided into three subdivisions of time: Early, Middle, and Late. Pennsylvanian rocks on the Colorado Plateau overlie Mississippian rocks, with a major unconformity or gap in the rock record between the two systems. For ten or more millions of years the underlying rocks, composed mostly of Mississippian Redwall and Leadville limestones, lay exposed to the agents of weathering and erosion. Chert, a rock type rich in silica that is often interbedded with limestone, eroded and left deposits of chert-pebble conglomerate that mark the unconformity. Soon after, as the landscape subsided or the sea level rose, shallow seas lapped across the region. A sequence of shallow marine, shoreline, and river deposits were laid down. They interfinger across the region as drab-colored marine rocks that gradually yielded to red-colored, continental-derived sediments.

The Weber Sandstone, a Pennsylvaniam coastal dune complex, is exposed along the Yampa River in northeastern Utah.

The oldest Pennsylvanian rocks are confined strictly to sedimentary basins and are not widespread. They contain red mudstone, sandstone, and tan limestone in the Grand Canyon's Watahomigi Formation. Adjacent parts of the plateau were slightly higher and became surfaces of erosion. As the Early Pennsylvanian continued, deposition became more widespread as sea level rose and connected the various sedimentary basins. Windblown sand also spread across much of the Colorado Plateau during this time and these deposits are preserved in the Grand Canyon region (Manakacha Formation), at Dinosaur National Monument (Weber Sandstone), and in many other places.

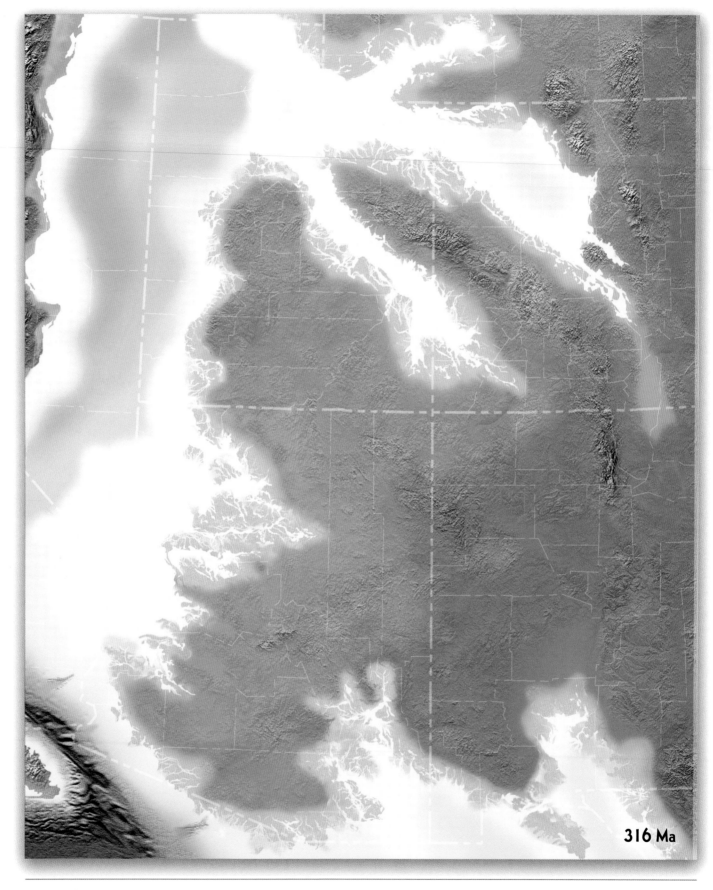

316 Ma

Above: Early Pennsylvanian paleogeography (316 Ma) marks a return of shallow seas to the Colorado Plateau region (Watahomogi Formation, Grand Canyon). The Ancestral Rocky Mountains began to rise and intervening areas began to subside.

Opposite: Early Pennsylvanian paleogeography at 312 Ma during deposition of the Manakacha Formation. In the early Pennsylvanian, paleotectonic and sedimentation patterns became clearly established and would persist throughout the Late Paleozoic. As the Ancestral Rockies rose, rivers and alluvial fans prograded into the lowlands. This map shows sedimentation patterns during a relatively low sea level.

312 Ma

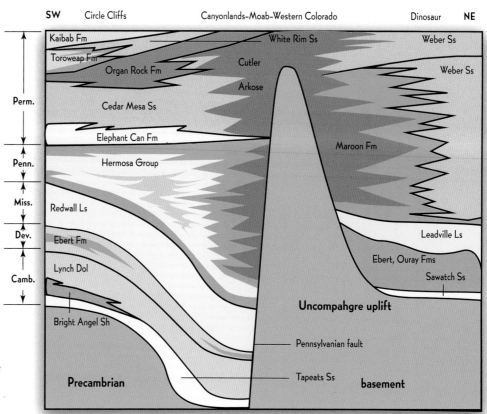

| SW | Circle Cliffs | Canyonlands–Moab–Western Colorado | Dinosaur | NE |

Right: Cross section restored to the end of Permian time showing how Pennsylvanian and Permian rocks thicken and thin across the tectonic elements. This diagram represents 4,000 to 10,000 feet (1,200–3,100 m) of sedimentary rock.

The close of the Early Pennsylvanian saw the Ancestral Rockies continue their long, pulsating rise. Streams originating in the highlands shed coarse gravel at the base of the mountain front. These gravels eventually formed conglomerate that was essentially left stranded in broad alluvial fans by the arid conditions at the mountain front. Their composition is rich in the Precambrian crystalline rocks exposed in the core of the Ancestral Rockies.

The Middle Pennsylvanian marks a time of coeval mountain uplift, basin subsidence, and high global sea level. These active conditions produced some of the more changeable and interesting deposits in the entire rock record of the plateau. Coarse conglomerates continued to be deposited in alluvial fans near the flanks of the Ancestral Rockies. However, a trough or basin developed southwest of and parallel to the mountain front, and an arm of the rising sea invaded this area. This trough has been given the name Paradox Basin, since evidence for its existence was first described from the Paradox Valley in southwestern Colorado. Carbonate rocks, fine-grained mudstone, salt, and gypsum were preserved in this linear basin, which stretched from central Utah to northwest New Mexico.

These variable rock types, collectively called the Paradox Formation, tell a fascinating story of how rapidly fluctuating sea levels affected deposition on this part of the plateau. When sea level was high, carbonate deposition dominated the outer fringes of this basin. Shallow-water creatures such as spirifer brachiopods and crinoids thrived in the clear, warm water, leaving their calcium-rich shells as limestone deposits. However,

Above: Cutler Formation (top layer) and Hermosa Group (bottom layer) near Durango, Colorado, along U.S. Highway 550

when sea levels fell, mudstone from the adjacent shoreline was washed over and covered these basin-fringing carbonates. Limestone deposition would then migrate to more central parts of the basin. Alternating bands of carbonate and mudstone can best be seen while floating through the canyons of the San Juan River in southeast Utah. Each cycle of sea level rising and falling lasted about 200,000 years, and these cycles compare favorably with the number of cycles observed in the recent glaciation (thirteen cycles in 2 to 3 million years).

Falling sea level had even more surprising results in the central part of the Paradox Basin. Here, restricted subbasins of briny seawater were left to evaporate in the hot desert sun. As the seawater dried out, it left deposits of gypsum first, then salt. Occasionally, small amounts of clastic debris from the Ancestral Rockies entered the fringes of these subbasins, leaving an interesting admixture of red continental mud, gypsum, and salt. These rocks can be seen in the depths of Cataract Canyon along the Colorado River (Canyonlands National Park), in the Spanish Valley near Moab, and in the Paradox Valley in Colorado. Their overall distribution documents where the Paradox Basin subsided most rapidly, since the briny, evaporating water would ultimately retreat to those areas of greatest subsidence. Sea level rise, controlled by cyclic glacial retreats, eventually started the cycle all over again—marine flooding leaving carbonate rocks first and then evaporation depositing gypsum and salt.

The accumulation of salt deposits had its own feedback mechanism which affected sedimentation within the depositional system. Salt behaves like putty when put under pressure, and as other sediments buried the salt, it was subjected to intense pressure. This caused the salt to flow upward, forming broad diapirs or salt domes precisely where salt accumulation was the thickest. A line of salt domes rose through the overlying deposits creating subtle topographic barriers in the central part of the Paradox Basin. This had the effect of keeping the coarse-grained gravel from the Ancestral Rockies confined to the eastern side of the basin—the clastic sediments simply could not overcome the rising barrier of salt domes.

The salt deposits have continued to migrate upward throughout much of subsequent geologic time and have "escaped" to the surface, where they are today partially dissolved and responsible for the presence of linear valleys in the modern landscape. These salt valleys have no obvious relationship to the rivers that flow across them, and it was for this very reason that the Paradox Valley was initially named; the Dolores River flows across, not within, the valley. The Colorado River does the same through the Spanish Valley, and the Salt Valley in Arches National Park is another location. Evidence for salt deformation within the rocks can also be seen in places along Utah Highway 128, northeast of Moab.

Pebble- and cobble-rich sediment types are an example of the coarse detritus that was shed into Pennsylvanian and Permian basins adjacent to the Ancestral Rocky Mountains. Shown is an example of the Cutler Formation at Fisher Towers, near Moab, Utah.

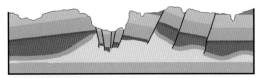

5. Late Neogene; folding and faulting as Colorado River water reaches buried salt

4. Early Paleogene; erosion and canyon cutting begin

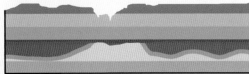

Cretaceous
Jurassic
Triassic

3. Triassic - Cretaceous deposition; salt deeply buried

Permian Cutler Fm

2. Permian deposition; weight causes salt deformation

Pennsylvanian salt

1. Pennsylvanian deposition; thick salt

Diagram showing the sequence of events that produced the salt diapirs or domes of western Colorado and eastern Utah.

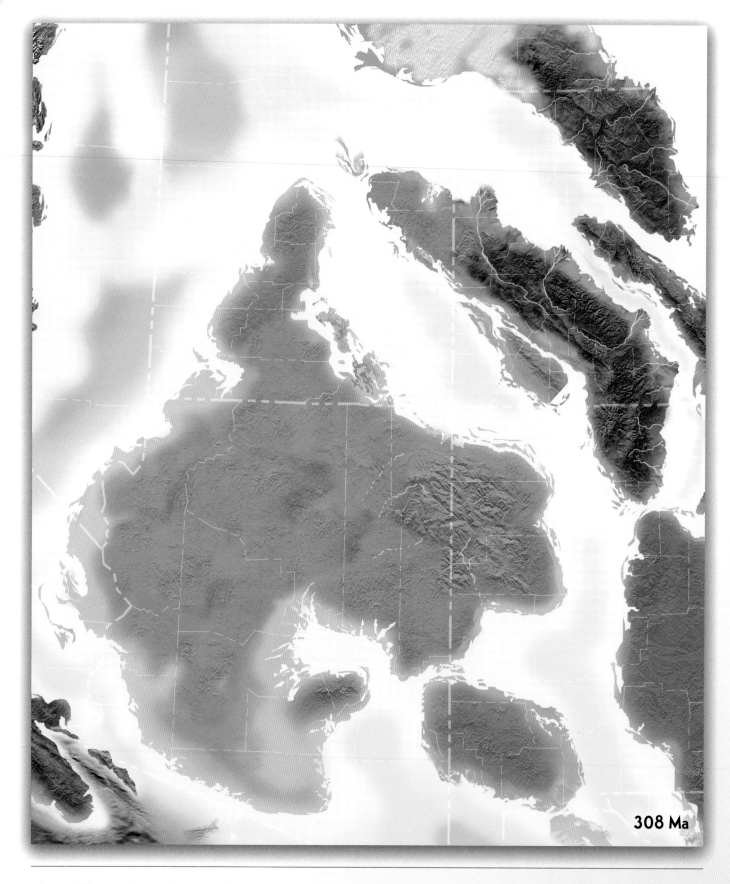

308 Ma

Above: Middle Pennsylvanian paleogeography (308 Ma). This map shows deposition during a sea level high. In southwestern Colorado and southeast Utah, normal marine deposits of the Paradox Formation were preserved.

Opposite: Middle Pennsylvanian time (308 Ma) during low sea level. The Paradox and Eagle basins became the sites of isolated seas (like the modern Caspian Sea), and large amounts of salt precipitated in the warm, restricted water. Calculations suggest a 400–600-foot difference between high and low sea levels, and the changes may have occurred rapidly in less than 200,000 years. As many as sixty cycles have been documented and resulted from fluctuating glaciations in the Southern Hemisphere.

308 Ma

304 Ma

Above: Late Middle Pennsylvanian paleogeography (304 Ma). Cyclic conditions caused by sea level fluctuations persisted, but salt deposition waned at this time.

Oppostite: Late Pennsylvanian paleogeography (300 Ma). Carbonate deposition was replaced by red sandstone and mudstone derived in continental settings. Sedimentation rates were likely quite high at this time as sediment-filled basins spilled laterally onto the broad shelf areas. As land areas expanded, eolian deposits spread across the coastal plains; these conditions left the Wescogame Formation in the Grand Canyon and the Honaker Trail Formation in southeast Utah.

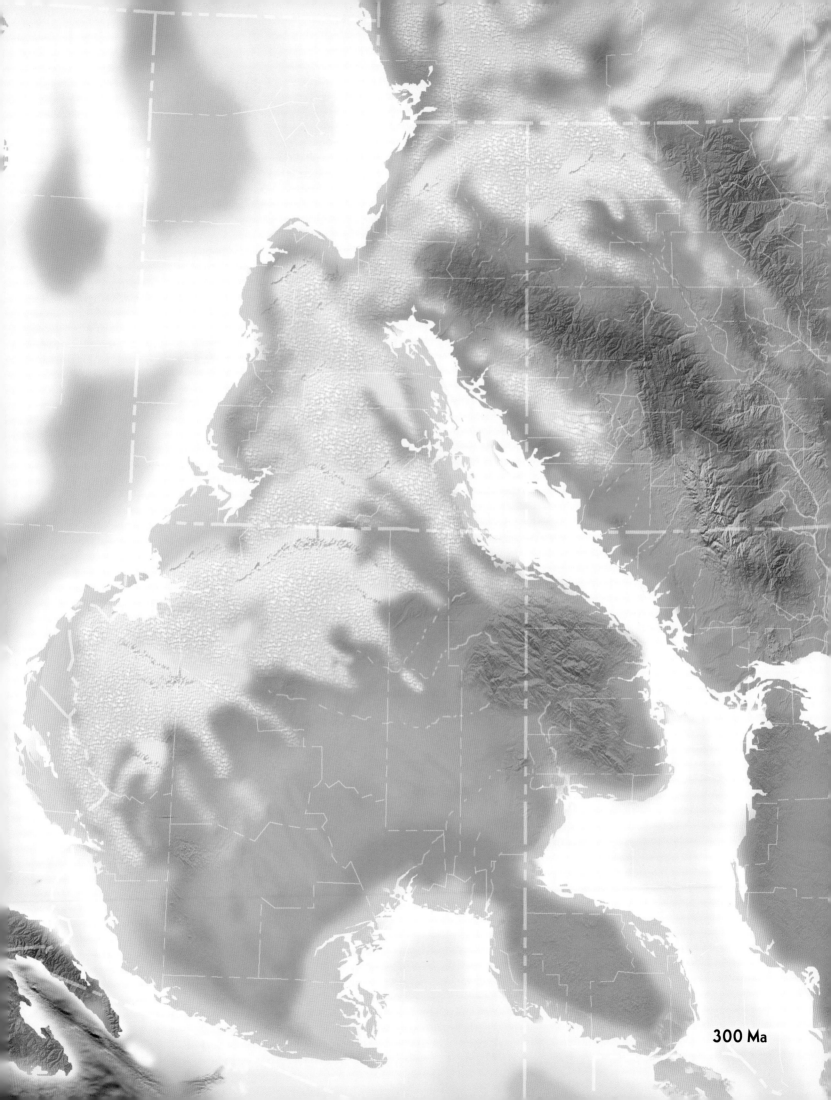

300 Ma

Above: Pennsylvanian and Permian rocks along the Colorado River below Dead Horse Point near Moab, Utah

Below: Fossils of Late Paleozoic marine fauna

The climate also affected depositional patterns at this time. The arid environment tended to strand deposits that were shed from the Ancestral Rockies near its mountainous edge. This happened because stream flow was ephemeral or even nonexistent and could not move the sediment very far. Consequently, the sea to the west was sediment-starved and carbonate deposition dominated during those times that were most arid. Other deposits are known from Middle Pennsylvanian basins adjacent to the plateau: the Oquirrh Group in the Oquirrh Basin in northern Utah; the Fountain Formation at the Garden of the Gods and Red Rock Amphitheater in Colorado; and the Horquilla Limestone in the Pedregosa Basin of southern Arizona.

The Late Pennsylvanian was a time that also saw widespread deposition across much of the region. This sedimentation became increasingly more complex as more windblown sand was introduced into the system, especially at Grand Canyon (Wescogame Formation), the Paradox Basin (Honaker Trail Formation), and northeast Utah (Weber Sandstone). Capping the upper 800 feet (244 m) of the San Juan River canyons and spectacularly displayed beneath the Goosenecks State Park overlook, the Honaker Trail Formation reveals a momentous transition in sediment styles from the dominantly marine conditions of the early and middle Paleozoic to the ubiquitous red-bed, continental deposits that would follow in the Permian, Triassic, and Jurassic. These few hundred feet of interbedded gray limestone, red mudstone, and red sandstone mark this great transition from dominantly marine deposits to dominantly continental deposits in the plateau's stratigraphy. As the Pennsylvanian came to a close 299 million years ago, carbonate deposition generally became less common and clastic sedimentation became more common, although cyclic deposition continued to dominate the entire region.

Desert Rivers, Dunes, and Seas: Permian

Some of the most strikingly beautiful modern landscapes on the Colorado Plateau are also a record of equally interesting landscapes from the Permian past. As Pangaea made its final assembly along the Appalachian front, it is perhaps interesting to note that one could have traveled completely overland from the Colorado Plateau all the way through Africa to Antarctica—Pangaea was at its greatest extent. The Permian was certainly a propitious time in the plateau region. Whether it be the stately towers of Monument Valley, the red-ochre rocks near Sedona, or the towering needles and natural bridges of the Canyonlands area, Permian rocks hold a fascination for both the modern plateau explorer and time travelers using their magic carpets of red strata!

View from Kanab Point at Grand Canyon. Permian rocks on the Colorado Plateau are some of the most colorful and interesting with respect to their origins.

Permian time is divided into three epochs: Early, Middle, and Late. On the Colorado Plateau, only rocks of the Early and Middle Permian were preserved; the Late Permian became a time of erosion or, more likely, a time of initial deposition with subsequent erosion (nonpreservation). Lower Permian rocks were deposited in conditions similar to those found in the Pennsylvanian, although the climate may have been a bit more arid. This is suggested by the abundant deposits of windblown sand that can be found from Montana to southeastern Arizona and from eastern Nevada to eastern Colorado. Lower Permian rocks are exposed at the surface or found subsurface in most of the Colorado Plateau region. An important exception is the area where the Uncompahgre uplift was located; it was still mountainous and continued to supply coarse gravel, sand, and mud onto adjacent shelves and basins to the west. Mountains cannot be the site of deposition since they are places where material is eroding away. In general, Lower Permian rocks on the Colorado Plateau reflect marine conditions in the west and central areas, and more terrestrial conditions closer to the mountain front to the east. This "ocean west–land east" pattern of deposition is a continuation of patterns established earlier in the Paleozoic.

An example of an early reptile making its way across a Permian sand dune

In the earliest part of the Permian a sea entered the region from the west and lapped back and forth repeatedly in transgression and regression across the landscape. This sloshing of the sea produced an interfingering of marine sediments with the nonmarine sediments coming from the highlands to the east. Much of the marine record for this time is dominated by carbonate rocks and many of these are fossiliferous and include species of brachiopods, corals, bryozoans, and other marine invertebrates. Marine carbonate deposits of the Lower Permian include the Pakoon Limestone in the western part of Grand Canyon National Park and the Elephant Canyon Formation or "lower Cutler beds" in Canyonlands National Park. (Note that some geologists suggest abandonment of the term Elephant Canyon Formation, but we include both names here pending final resolution of the dispute.)

290 Ma

Above: Early Permian paleogeography (290 Ma). During this interval, a dynamic interplay between continental and shallow marine environments is observed in the rocks in Utah and Arizona. Fluvial, eolian, beach, tidal flat, and marine-shelf environments were in close proximity and subtle but rapid shifts in sea level generated complex interbedded deposits. As the seas danced back and forth, the Elephant Canyon Formation was deposited in Utah, while the Pakoon Limestone was deposited in Arizona.

Opposite: Early Permian paleogeography of the Colorado Plateau at 287 Ma. During this time, the Cedar Mesa Sandstone was deposited in Utah and the Esplanade Sandstone in the Grand Canyon. Although the source of these sands has yet to be adequately documented, the paleogeography suggests that appreciable amounts were derived from river systems that drained the Appalachians and the nearby Ancestral Rockies.

287 Ma

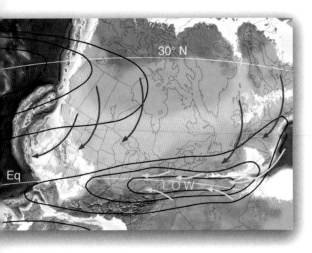

Postulated paleoclimate of the North American portion of the Pangean landmass. The Colorado Plateau region most likely experienced strong, consistent, dry trade winds from the northeast (ancient direction).

Permian eolian sandstone exposed at Spider Rock in Arizona's Canyon de Chelly National Monument

The terrestrial rock record of the Lower Permian includes interbedded eolian (from Aeolus, Greek goddess of wind) sandstone and fluvial red sandstone and mudstone. Conglomerate is abundant near the Ancestral Rockies uplift and is known as the Cutler Formation. Coeval eolian deposits include the Cedar Mesa Sandstone in Canyonlands National Park and Natural Bridges National Monument, and the Esplanade Sandstone in the Grand Canyon. These eolian sands spread southward on the northerly winds and accumulated in great windblown dune fields called ergs (Arabic for sea of sand). These dunes, once perhaps hundreds of feet high, had south-dipping slipfaces that formed inclined laminae called cross-beds. Eolian sand both avalanched down and was blown across these slipfaces, causing the dunes to migrate downwind. This process left a surprising and fortuitously rich record of wind direction from this pre-weatherman ancient time.

After deposition of the eolian Cedar Mesa and Esplanade sandstones, fluvial redbeds from the Ancestral Rockies covered them. These rocks are called the Organ Rock Formation (Canyonlands National Park and Monument Valley) and the Hermit Formation (Grand Canyon and Sedona). Both formations are coeval and brick red in color. They formed when major arid-land river systems spread sediment southwest from the Uncompahgre uplift across much of the Colorado Plateau; all previous basins had been overwhelmed and filled with sediment by this time. River channel deposits are composed of sandstone and conglomerate, documenting where the flow is strongest and moves the coarsest deposits. Mud and fine sand were deposited on the floodplains, defined as those areas that were under water only during large floods. These strata become fine-grained toward the top, reflecting the waning stages of the floods and the decreasing flow energy.

The close of the Early Permian saw an even greater influx of windblown sand into the region. Dunes deposited the De Chelly Sandstone in northeast Arizona (Monument Valley and Canyon de Chelly National Monument) and the Schnebly Hill Formation in north-central Arizona (Sedona). The Holbrook Basin was the site of particularly rapid subsidence and resulted in the increased preservation of this sandy sediment. Northern winds blew the sand toward this basin, and rapid subsidence allowed for the gradual incursion of the Pedregosa Sea from the southeast. This kept the Sedona area very close to sea level. Even though some deposits originated as dunes in eolian environments, they ultimately came to rest as tidally reworked sediments, known as the Bell Rock Member of the Schnebly Hill Formation (members are further subdivisions of formations).

Eventually, the Pedregosa Sea transgressed all the way into the Sedona area and left shallow marine deposits known as the Fort Apache Member. The Fort Apache is only about eight feet (2.5 m) thick near Sedona, but pinches out just west of there. It thickens southeast to more than 100 feet (30 m) near Fort Apache, Arizona. These relationships

Left: Global view of Pangaea during the Permian Period, 260 Ma

Below: The present is often used to understand the past. A modern dune field at Ackchar, Mauritania, in West Africa, exemplifies the Permian desert landscape of the Colorado Plateau.

260 Ma

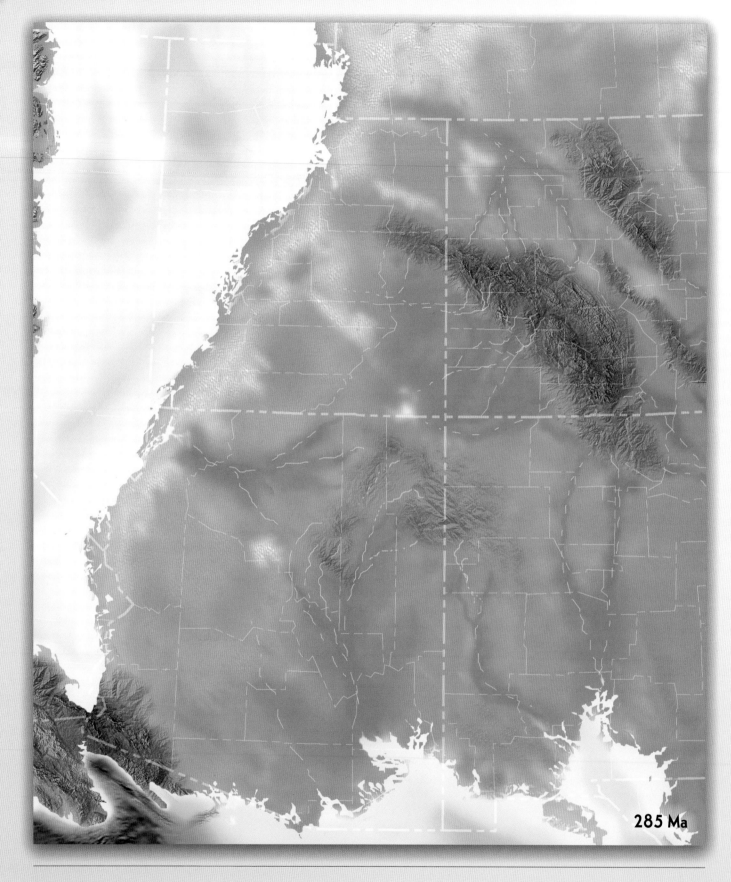

285 Ma

Above: Paleogeography 285 Ma showing that Early Permian dune conditions gave way to more fluvial influences. Redbeds are common in the Hermit, Abo, and Organ Rock formations, which display abundant channel-shaped sandstone bodies and mudstone beds, revealing stream deposition in channels and floodplains respectively.

Opposite: Paleogeography 280 Ma. Deposits for this time slice are found only in basins in western New Mexico and northeastern Arizona (the Holbrook Basin), and they produced fine-grained restricted marine or sabkha redbeds. No deposits were preserved in southeast Utah or the Grand Canyon. Rimming this basin to the north and west were eolian dune fields—the De Chelly and Schnebly Hill formations.

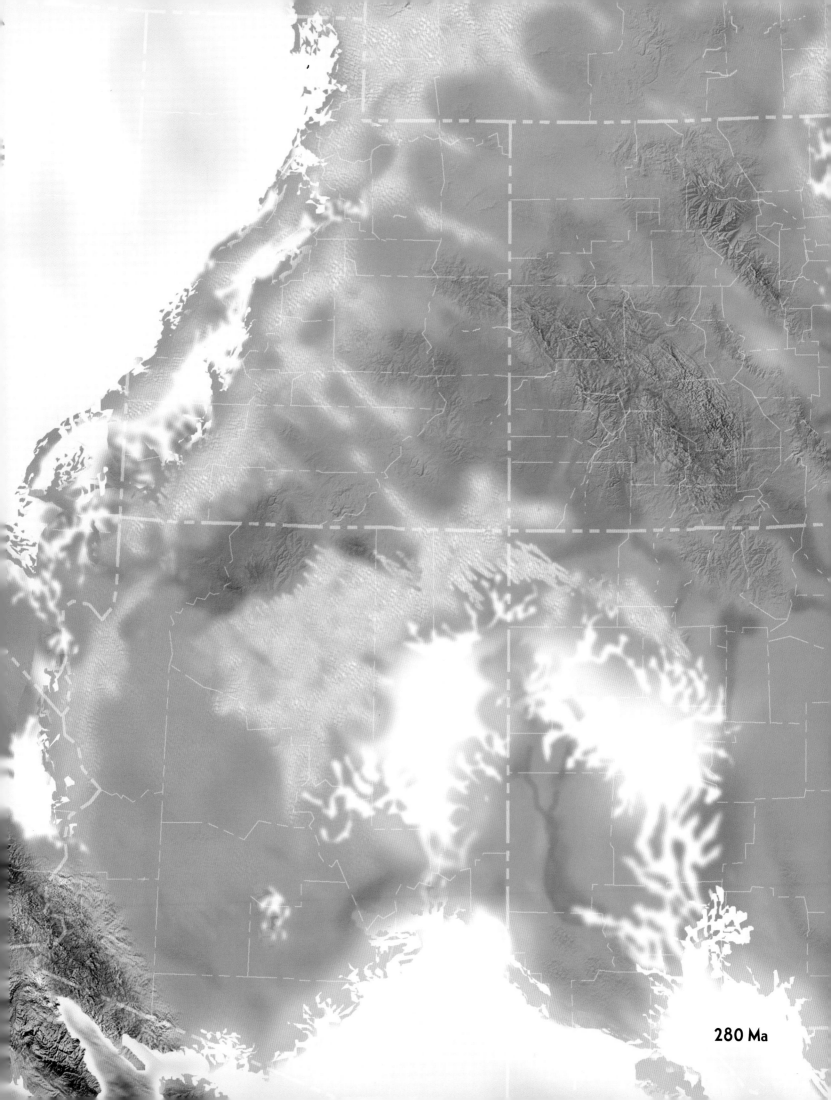

280 Ma

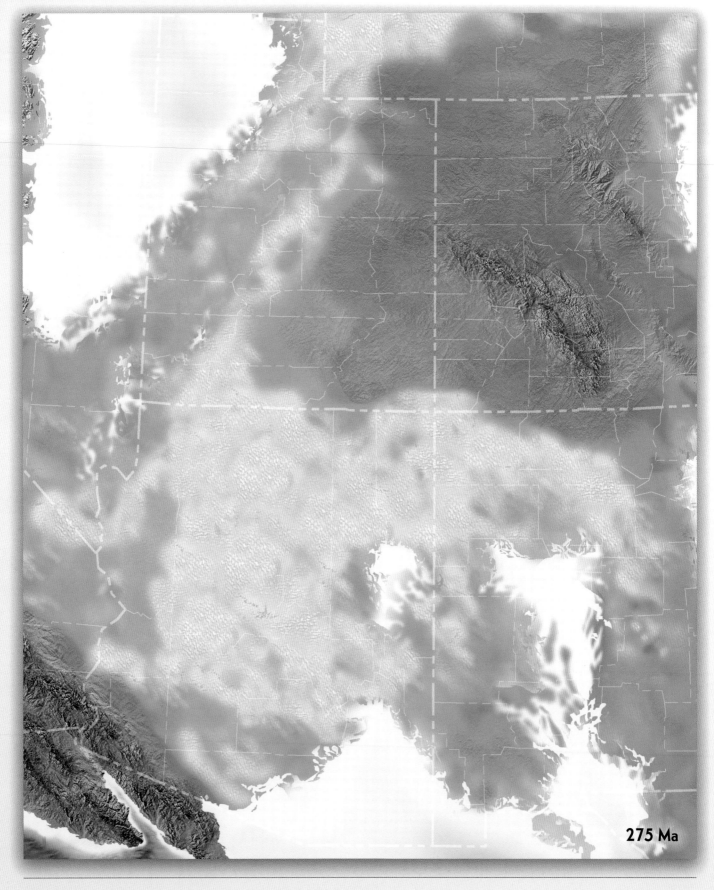

275 Ma

Above: Paleogeography 275 Ma when the Coconino Sandstone accumulated in a great eolian desert. This covered most of northern Arizona and northern New Mexico (as the Glorieta Sandstone), but did not extend north into Utah or Colorado.

Opposite: Early Permian paleogeography at 272 Ma, when marine environments encroached from the west. The Toroweap Formation preserves coastal and shallow marine deposits in western Utah and Arizona, while the White Rim and upper Coconino sandstones were preserved onshore as dunes along this encroaching sea.

272 Ma

record that the Pedregosa Sea was present in the Sedona area for only a short period of time and more long-lived, and at times deeper, to the southeast. The purely eolian Sycamore Pass Member overlies the Fort Apache Member and shows how sandy dunes eventually overwhelmed the Pedregosa Sea in the Sedona area.

The Fort Apache Limestone Member (foreground) of the Schnebly Hill Formation near Sedona, Arizona, marks the northward advance of the Pedregosa Sea.

Some portions of both the De Chelly Sandstone and the Schnebly Hill Formation formed under sabkha conditions. Sabkha is an Arabic word referring to places in arid regions that have high water tables. They are literally desert swamps. A shallow water table in a desert setting brings water to the surface in a reflux pumping action—as the water evaporates it pulls more groundwater upward. This reflux process leaves crusty minerals such as calcite, dolomite, gypsum, and salt in the soil just below the surface. These salts push away existing sediment as they grow, distorting original bedding into a crinkly and contorted mass. This bedding pattern is a good indicator of sabkha conditions in the past and is common in Permian rocks across the southern Colorado Plateau region.

During deposition of the De Chelly and Schnebly Hill formations, sand and sabkha deposits were preserved only where the earth's crust was actively subsiding. Nearby areas that were stable and not subsiding did not preserve sediment. These stable shelves acted merely as surfaces of transport for the sand grains and dune fields. This explains why there is no Schnebly Hill Formation or De Chelly Sandstone in areas such as the Grand Canyon today. Although the dunes most likely were present there and sand did blow through the area, it simply did not have the accommodation space that placed the sediment into storage. Sediments can only be preserved when there is accommodation space to store them. Many a Grand Canyon hiker has graciously thanked the stratigraphic gods that those extra 700 feet (213 m) of Schnebly Hill Formation were not preserved in an abyss already plenty deep!

With time, the Pedregosa Sea retreated far to the southeast and desert dune conditions spread to all of northern Arizona. The Coconino Sandstone was deposited in a large erg that at one time may have covered almost all of Arizona and much of western New Mexico, where it is called the Glorieta Sandstone. Four- and five-toed reptiles walked across these Coconino dunes leaving their footprints on the angled cross-beds. Spectacular examples abound on some Grand Canyon trails, especially the Hermit Trail.

Toward the end of the Early Permian, two marine transgressions finally completed the plateau's Paleozoic section of rocks. Both seas entered the region, this time from the west. The first of these transgressions deposited marine, sabkha, and shoreline sediments known as the Toroweap Formation. The shallow marine and sabkha conditions extended east only to a line drawn roughly between Sedona and the Lees Ferry area in Marble Canyon. In these western outcrops, composed of fine-grained siltstone and gypsum, the Toroweap forms slopes of easily eroded material (Grand Canyon). However, to the east of that line, eolian shoreline environments deposited resistant cross-bedded sandstone that laterally replaces the softer sabkha deposits. These eolian deposits form an "upper Coconino Sandstone" (or "sandy Toroweap Formation," take your pick) in northern Arizona (Oak Creek Canyon and Walnut Canyon National Monument) and the coeval White Rim Sandstone (Canyonlands National Park) in southeast Utah. This curious horizontal change in lithology is called a facies change (change in character or rock type) and provides the critical, detailed information needed to interpret how the geography changed from one place to the next at a single time.

The second marine transgression produced the widespread Kaibab Formation. This transgression reached farther east than the Toroweap and covered the western half

Below: A spectacular reptile trackway in the Coconino Sandstone along the Hermit Trail in Grand Canyon National Park. Alternating left-to-right footprints can be seen with a "tail drag" between them.

and southern two-thirds of the Colorado Plateau with marine carbonate and sandstone. Brachiopods, mollusks, bryozoans, corals, and sponges are all common in the Kaibab. This unit marks the top of the Permian section of rocks and forms the rim of Grand Canyon and much of the Mogollon Rim. The upper parts of the Kaibab are Middle Permian in age—the only deposits of this age on the plateau.

Following the last Kaibab transgression, seas retreated from the entire region and an unconformity developed on top of the Kaibab Formation. Late Permian rocks are not present on the Colorado Plateau except perhaps in northern Utah. On the plateau, and almost everywhere on Earth for that matter, several tens of millions of years in the rock record are missing. During this episode of earth history, sea level may have been at an all-time low. This kept the marine environment restricted to the deeper and light-poor ocean basins, where shallow-water creatures could not live. This condition created havoc for these shallow marine organisms as most of their territory was lost. The ensuing extinction of marine life, perhaps attributed to the loss of shallow marine environments worldwide, was the greatest in the entire rock record; more than 90 percent of all species went extinct. Another possible factor precipitating these extinctions is significant climate change caused by meteor impacts or volcanic eruptions.

Top: Both the Kaibab and Toroweap seas were bordered with continental sand dunes similar to these on the coast of Namibia in southern Africa.

Inset: A typical fossil brachiopod from the Kaibab Formation in the Grand Canyon/Sedona region

---

Opposite: Middle Permian paleogeography 270 Ma. The Kaibab Formation reflects similar depositional conditions to the underlying Toroweap Formation but spreads farther east. Kaibab-age seas covered much of the Southwest from southern New Mexico to Nevada and represent the last of the great Paleozoic seas of North America.

270 Ma

The late Paleozoic marks a time when significant parts of the stratigraphic section across parts of the Western Interior were dominated by continental deposits. As with marine deposits, space is necessary for the preservation of sediment that originates in continental settings. In this instance, it was perhaps an overabundance of sediment, rather than subsidence that was more important in the preservation of this sediment. In other words, sediment accumulated and expanded at a rate faster than space was created, which allowed continental environments (ergs and rivers) to expand across areas that were once occupied by shallow marine shelves; the sediment fill occupies volume so the seas were forced out of the region. These continental environments remained above sea level, though just barely. Minor fluctuations in the delicate balance would allow for the return of shallow seas. In some places like the western Grand Canyon, conditions fluctuated rapidly and repeatedly to produce complex interbedded sequences of marine and nonmarine rocks. Glaciers at the South Pole lasted well into the Permian.

Locations of Pennsylvanian and Permian Rocks

Pennsylvanian rocks are well-exposed in the Grand Canyon and Sedona regions where the lower two-thirds of the Supai Group make up rocks of this time. They are also exposed along the San Juan River at Goosenecks State Park and the Colorado River at Canyonlands National Park in southeast Utah. These rocks can also be seen along U.S. Highway 550 north of Durango, Colorado, in the Hermosa Cliffs, which is the type section of the Hermosa Group (a package of three formations containing the Pinkerton Trail, Paradox, and Honaker Trail formations).

Permian rocks are widely exposed on the Colorado Plateau. They comprise roughly the upper one-third of the walls of Grand Canyon and much of upper Marble Canyon; they form the spectacular scenery at Sedona and Canyon de Chelly. The "monuments" in Monument Valley along the Utah-Arizona border are formed from Permian rocks. In southeast Utah, they comprise the ramparts of Cedar Mesa and the pinnacles of Valley of the Gods; they form the stone bridges of Natural Bridges National Monument and the surrounding scenery of White Canyon. Permian rocks are responsible for the spectacular rock forms in the Needles, Land of Standing Rocks, White Rim, Monument Basin, and Shafer Basin in Canyonlands National Park. They cap the spectacular Fisher Towers east of Moab and are host to the slot canyons in the central San Rafael Swell near Capitol Reef National Park. Finally, they also form much of the canyon walls of the Green and Yampa rivers in Dinosaur National Monument, Utah-Colorado, and the centers of the salt anticlines in western Colorado.

The final Paleozoic deposit on the Colorado Plateau is the Kaibab Formation, seen here at Duck on the Rock, Grand Canyon National Park.

Chapter Summary

The Late Paleozoic Era saw the final amalgamation of the Pangaean supercontinent and ushered in a long period of continentally dominated sedimentary sequences on the Colorado Plateau. An important landscape element on the plateau at this time was the rise of the Ancestral Rocky Mountains, uplifted as a result of the viselike compression accompanying continental amalgamation. Repeated sea level fluctuations, arising from glacial conditions in the Southern Hemisphere, resulted in over sixty cycles of sedimentation during the Pennsylvanian Period and included the deposition of significant salt in the Paradox Basin. As the Permian began, these fluctuations were less of an influence here because of the increased amount of eolian and fluvial depositional settings. A great extinction and unconformity occurred at the end of the Paleozoic, setting the stage for another round of continental deposits as the Mesozoic Era began.

Below: The Cedar Mesa Sandstone, exposed in the Needles District in Canyonlands National Park, Utah

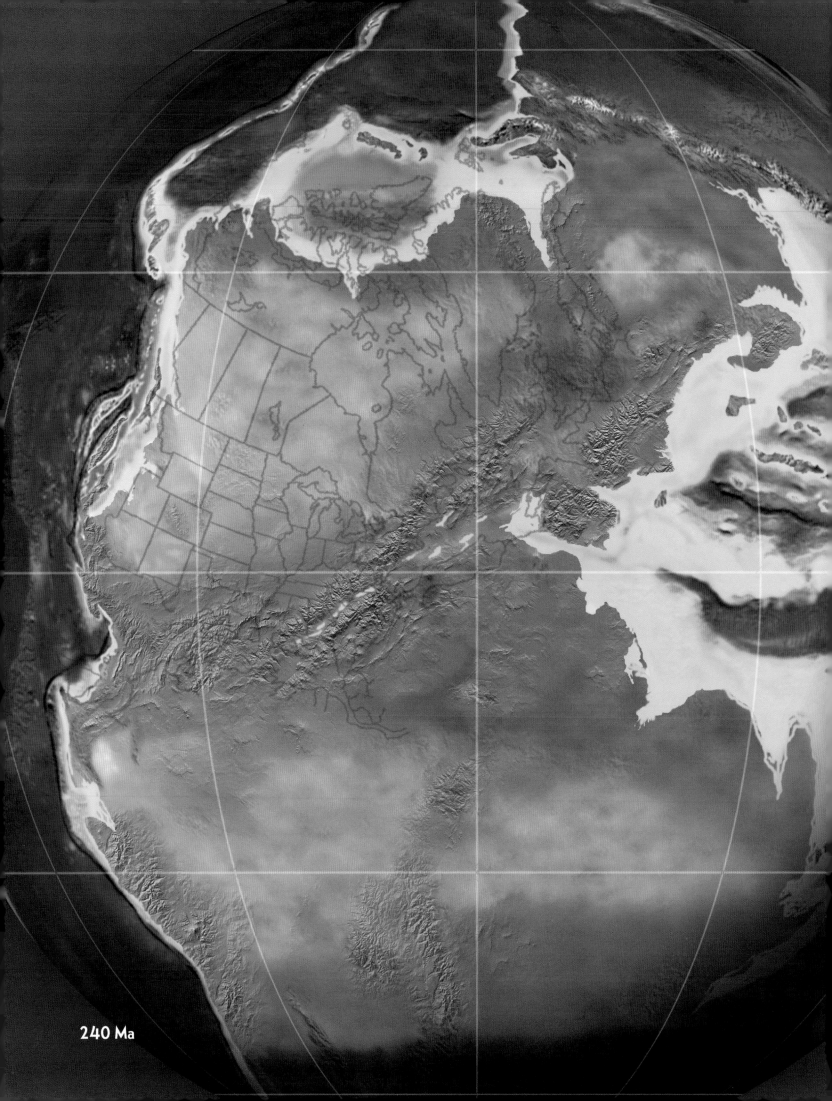
240 Ma

CHAPTER FIVE
DESERT RIVERS AND GREAT DUNES
TRIASSIC AND JURASSIC: 251 TO 145 MILLION YEARS AGO

At the dawn of the Mesozoic Era, before the rise of the great dinosaurs, all of the earth's continents were assembled into a giant supercontinent called Pangaea. This huge landmass stood high and dry—even parts of the usually sea-bound continental shelves were exposed to coastal erosion. These fringes of the continent, oftentimes the site of limestone deposition during the Paleozoic, were now arid, low-lying coastal plains where weathering and erosion dominated. This was a time when deposition and preservation on the Colorado Plateau took a 25- to 30-million-year hiatus.

One possible reason for this significant break in sedimentation may relate to the very presence of a single, large continental mass like Pangaea. Continents have relatively thick crusts (about twenty to forty miles/32 to 64 km) compared to ocean crust (at only about six miles/10 km thick). Thus, continents act like giant insulating blankets that can capture and hold the upward flow of heat generated within the earth's core. Continents, especially large ones, become places where heat accumulates. This causes continental rocks to become less dense, making them more buoyant and causing continents to become elevated. A rather small increase in temperature within a continent can result in a measurable gain in elevation. As continents rise, seawater is displaced from shallow continental shelves, exposing them to periods of erosion or at least nondeposition. Relatively warm, high continents are thus the sites of weathering and erosion, whereas low, cooler continents are the sites of subsidence and deposition.

170 Ma

As the Early Triassic commenced, parts of western North America once again began to subside. This subsidence was related to subduction, a process where ocean crust is pushed beneath the edge of a continent, becoming partially consumed and melted as it descends into the earth's mantle. The ultimate cause of this subduction in western North America originated two

Opposite: Global view of North America in the Triassic. Pangaea shows early signs of rifting between Africa and North America (where rift basins are shown), while subduction becomes established along the western margin of North America.

Above: Global view of North America in the Jurassic. Note the narrow Atlantic Ocean, just beginning to open up. The shapes of North America, Africa, and South America are evident.

Fluvial deposits of the Moenkopi Formation north of Winslow, Arizona

Above: Modern ripple marks on a tidal flat at Rocky Point, Mexico

Below: Ancient ripple marks from the Moenkopi Formation near Cameron, Arizona

thousand miles (3,200 km) east on the far side of the Appalachian chain. As trapped heat caused the central portions of Pangaea to uplift and rift, North America began to slide westward, separating itself from Africa and Europe. This initiated the shaping of the modern continents and began the long-lived subduction environment in the far western portions of North America. On the Colorado Plateau, this subduction merely pulled at its edge, causing it to subside gently. Subsidence rates were greater in central Nevada than on the plateau and resulted in Mesozoic rocks that become progressively thicker toward the west in this region.

Great River Systems of the Moenkopi and Chinle: Triassic Deposition

The greater bulk of Triassic-age sediments on the Colorado Plateau originated in fluvial (river) settings. During this time, the plateau region was a uniform and somewhat monotonous coastal plain that stretched from Texas and New Mexico, across Arizona and into central Utah. The plain was flat-lying to such a degree that when the sea level rose slightly, great swaths of the coastal plain were inundated with seawater. When sea level dropped, a huge expanse of exposed land appeared. It was upon this setting that the Moenkopi Formation was deposited as interbedded fluvial and tidal sediments. But where did the rivers originate?

Eastern North America at this time was quite mountainous and the continental divide ran along the crest of the Appalachian chain from New England to central Mexico (the Gulf of Mexico did not yet exist and portions of modern Mexico were once more closely connected to the southern United States). Rivers that originated near this far-off divide flowed westward across the entire continent to their ultimate destiny on the coastal plain. The Colorado Plateau region was probably arid, perhaps even hyper-arid, and only the largest of the mountain-born rivers made it all the way to the sea, then located in modern-day Nevada. The rivers' low gradient resulted in fairly sluggish flow that deposited mostly mudstone and very fine-grained sandstone. Near the fluctuating shoreline, broad tidal flats covered vast areas. The extremely low gradient exaggerated the distance traversed by daily tides, although the exact value of either tidal range or distance traveled is not known. Marine deposits to the west consisted of gray mudstone and tan limestone. During several periods of relatively high sea level, marine deposits extended southeastward as thin tongues of gray rock, which became interbedded in sharp contrast to the dominant red sandstone and mudstone of the continental Moenkopi Formation.

The Moenkopi is known for its ubiquitous and well-preserved ripple marks and mud cracks. These sedimentary structures formed both on river floodplains and coastal tidal flats. Ripples result when shallow water flows across a noncohesive bed such as gravel, sand, or silt (mud is an example of a cohesive bed). The force of flowing water tugging at the clastic debris produces the ripple marks. Wind-whipped waves moving across very shallow water can also produce ripple marks. These features commonly form on tidal flats, river floodplains, and even some channel bottoms, all of which are well represented in the Moenkopi Formation. Mud cracks form when cohesive sediment such as mud or clay dries out and contracts under the scorching sun. These cracks usually display polygonal shapes and result from the shrinkage of the sediment volume upon drying. These conditions commonly occur on floodplains, in shallow lakes and ponds, and along the upper reaches of tidal flats, again, environments well represented in the Moenkopi Formation (Capitol Reef National Park and Wupatki National Monument).

Triassic Moenkopi Formation (lower ledges and slopes), Chinle Formation (upper ledges and slopes), and Jurassic Wingate Sandstone (upper cliffs), eastern San Rafael Swell, Utah

In the Early Triassic, beginning with deposition of the Moenkopi Formation, a drainage configuration developed in the plateau region that would generally persist for the next 70 million years. It was oriented diagonally across Arizona and Utah and, using modern coordinates, trended from the south and east toward the north and west. Although these northwest-directed rivers would become overwhelmed and buried in windblown sand at various times, the return of wetter conditions usually reestablished this particular drainage pattern. The die was cast sometime in the Early Triassic for this long-standing drainage configuration, which lasted well into the Jurassic.

Badlands within the Moenkopi Formation overlain by a bench of Chinle Formation. Cliffs of Navajo Sandstone at Zion National Park form the background.

The Moenkopi can contain localized fossils, but most exposures are usually devoid of fossils. Reptiles and amphibians lived along the streams and lakes of this pre-dinosaur landscape, and their trackways are sometimes encountered. Actual bone material is much rarer in the Moenkopi. Many floodplain deposits are marked by the traces and trackways of invertebrates that crawled and burrowed into the soft mud. Marine deposits contain mollusks, the most interesting of which are the ammonites, similar in form to the modern nautilus and useful for the detailed correlation of rocks of this age.

It was during the Middle Triassic that Moenkopi deposition came to an end as the Colorado Plateau region once again was subject to weathering and erosion. An unconformity at the top of the formation marks this depositional break but the

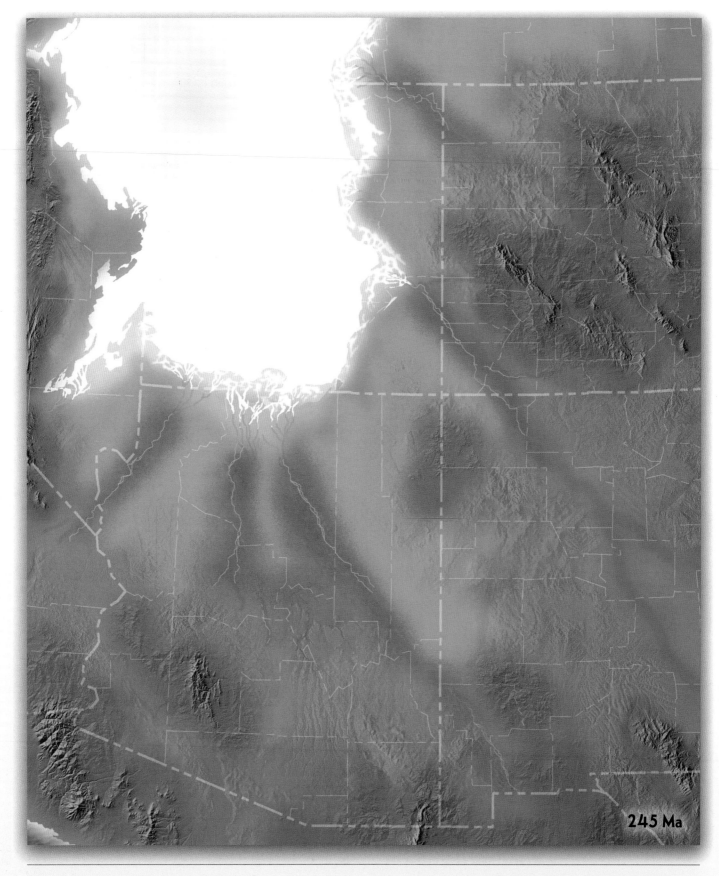

245 Ma

Above: Early Triassic paleogeography (245 Ma). After as many as 25 million years of erosion, subsidence resumed and the Moenkopi Formation covered the region. This unit is rich in fluvial and tidal redbeds to the south and east, while shallow seas from the northwest deposited the Timpoweap and Sinbad members there. The Virgin Limestone member records a slightly later incursion of the sea in the Lower Triassic.

Opposite: Middle Triassic paleography (240 Ma). The Triassic seas retreated to the northwest, and vast, arid rivers flowed from the southeast. Reptiles and amphibians flourished along the river systems that gave rise to the Holbrook Member of the Moenkopi Formation.

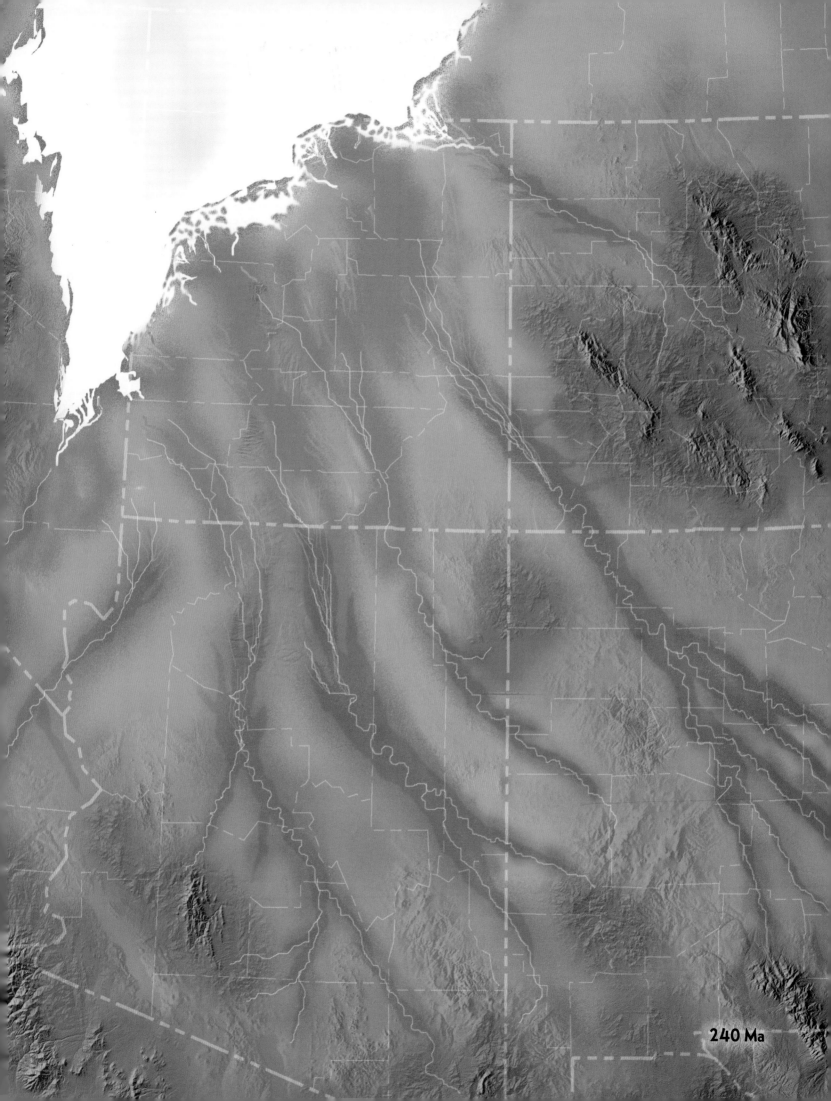

240 Ma

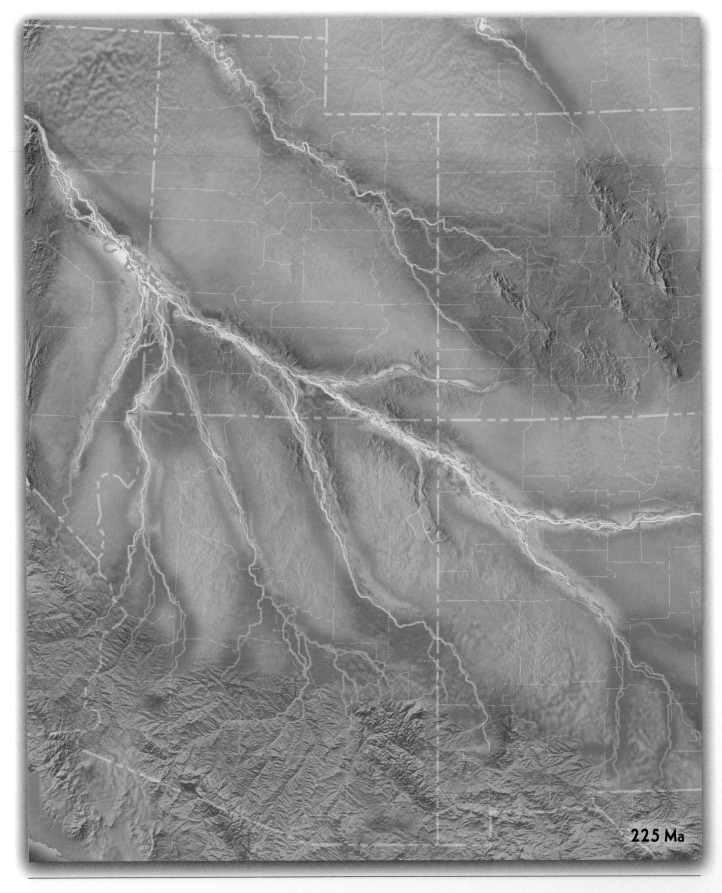

225 Ma

Above: The Late Triassic paleogeography (225 Ma), when the Shinarump Member of the Chinle Formation filled valleys cut into the Moenkopi Formation. Analyses of the coarse sandstone and conglomerate suggest that these braided stream deposits were from big rivers—tens of feet deep and hundreds of meters wide.

Opposite: Late Triassic landscape at 215 Ma for the Petrified Forest Member of the Chinle Formation. These rivers were large and meandering with their headwaters in the uplands of southern and eastern North America (Appalachians). Cutoff meanders called oxbow lakes lie adjacent to the rivers, and these swamps were the perfect locale for volcanic ash (from the south and west) to be preserved.

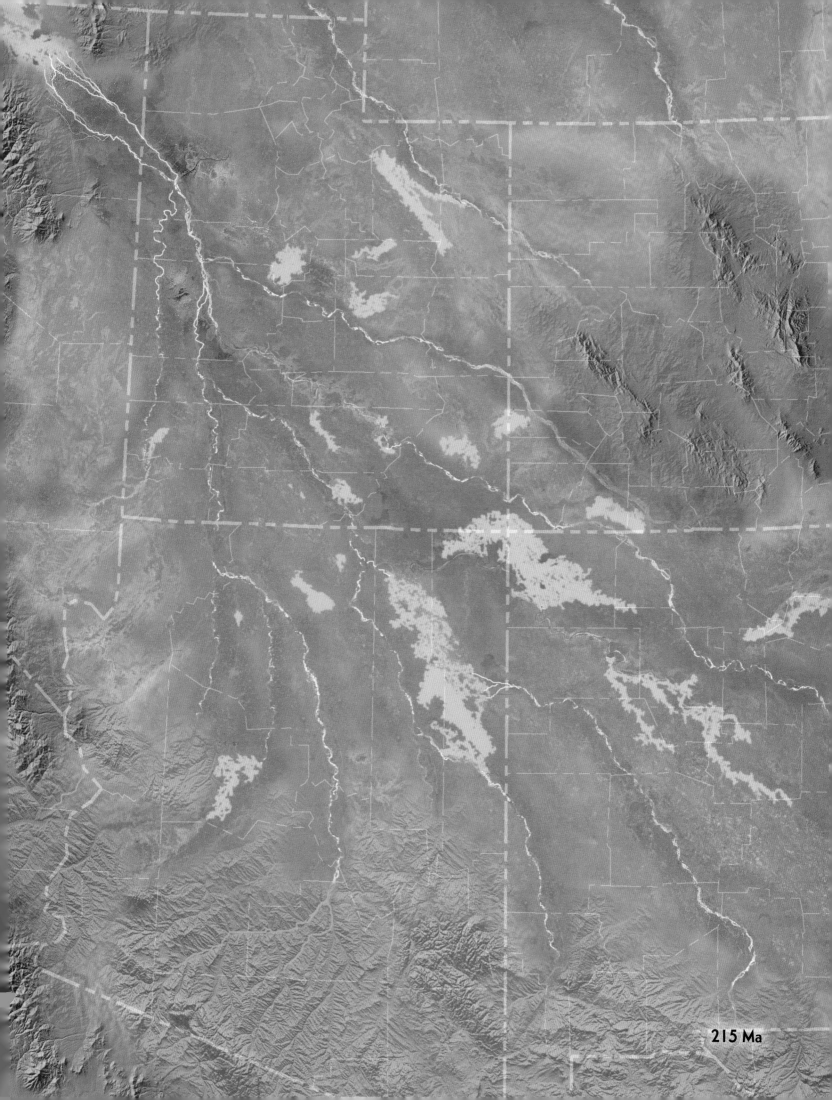

215 Ma

precise cause for it is unknown. Possibilities include a change in climate and the resulting change in the style and size of rivers, a pronounced drop in sea level, or a slight uplift of the Colorado Plateau region. Whatever the reason, the rather soft mudstone and sandstone beds of the Moenkopi Formation were partially eroded during this time. Rivers incised the landscape and produced relief that ranged up to several hundred feet in some places.

A petrified log, belonging to the extinct genus Araucarioxylon, *found in Petrified Forest National Park, Arizona*

In the Late Triassic, however, the Colorado Plateau region began to subside once again. While perhaps becoming somewhat repetitive to note, the gentle, alternating rise and fall of the ancient landscape through time can be viewed as analogous to the repetitive inhalations and exhalations of a sleeping giant, whose chest rises and falls with each breath. The surface of the future Colorado Plateau certainly behaved in such a fashion throughout its early Mesozoic history—each "breath in" causing the land to gently rise and each "breath out" causing it to fall.

Stream deposits filled the valleys that had been cut into the top of the Moenkopi Formation at the start of the Late Triassic. This sediment was a very coarse-grained sandstone and pebble conglomerate that historically has been called the Shinarump Conglomerate. Today its official name is the Shinarump Member of the Chinle Formation. Shinarump gravel initially filled only the lowest valleys, but through time it spread more widely across the broader Moenkopi surface. The deposit is absent, however, where Moenkopi remnants stood too high to allow for the gravel to accumulate.

The Andes Mountains in South America are a perfect modern analogy for the Triassic landscapes in Arizona. Uplands with active volcanoes were present in the Mogollon Highlands to the south, and trees similar to the Araucaria *species grew on the Chinle floodplans. This photograph was taken near Lonquimay, Chile.*

Because of the relief on which the Shinarump was deposited, its thickness varies widely from zero, to a few feet, to more than two hundred feet (60 m) thick. Shinarump streams were apparently more vigorous than those of the underlying Moenkopi. Clasts are commonly several inches across and consist of durable rock types that were eroded from the uplands hundreds of miles away to the south and east. However, even more surprising details have recently been unearthed. The discovery of the mineral zircon within some deposits of the Chinle Formation has allowed geologists to better understand the larger setting of this river system. Zircon is a highly durable, igneous mineral that can travel great distances in rivers. It contains a specific chemical signature that can reveal its source area, providing that the source rocks are not completely gone from the modern landscape. Grains of zircon found in the Chinle Formation on the Colorado Plateau have an identical chemical signature to zircons found in granites still in place in the southern Appalachians, almost a thousand miles (1,600 km) away. River systems of this magnitude compare favorably with some parts of the Amazon River today; the only exception being the humid modern climate of the Amazon and the semiarid setting of the Triassic on the plateau. This zircon evidence is quite recent and allows for more precise reconstruction of the early Mesozoic landscape.

An outcrop of the Petrified Forest Member of the Chinle Formation near Cameron, Arizona

Rivers continued to dominate depositional patterns during much of the remainder of Chinle time. However, a much higher proportion of mudstone was preserved during the accumulation of overlying units, such as the Petrified Forest Member of the Chinle. Mudstones occur in unusual and riotous shades of red, gray, brown, purple, tan, orange, and pink. They formed as overbank deposits when floodwater escaped the confines of a river channel and deposited widespread mud across the featureless floodplain. Occasional lenses of sandstone and conglomerate are interbedded with these soft mudstones and define the specific locations of the sinuous river channels. Today these thin sandstones and conglomerates cap the mesa tops in many a Painted Desert landscape in Petrified Forest and Capitol Reef national parks.

A plethora of plants and animals lived in and along the Chinle rivers. This was the time, about 225 million years ago, when dinosaurs made their first appearance in North America. Coelophysis had the dimensions of a large-sized dog, and spectacular remains have been found at Ghost Ranch in New Mexico and Petrified Forest (specimens may be seen at the Museum of Northern Arizona in Flagstaff). Giant phytosaurs with rows of menacing teeth similar to those of a large crocodilian, roamed the Chinle rivers and preyed on anything that crossed their path. Abundant petrified wood, much of it found as large, colorful tree trunks lying on their sides, weather out from these rocks. The petrified logs are the namesake of the Petrified Forest Member of the Chinle Formation and belong to the genera *Araucarioxylon* and *Woodworthia*. (*Araucarioxylon* initially was named because it was thought to be related to araucaria trees currently found in the southern hemisphere; *Araucarioxylon* is now thought to be a primitive conifer and not related to the *araucarias*.) Less spectacular but equally important are the invertebrates and other smaller plant fossils.

Triassic sediments on the Colorado Plateau are dominated by fluvial deposits like these in the Chinle Formation near Fry Canyon, Utah. The Jurassic Wingate Sandstone forms the cliffs above.

It is important to note one other aspect of the Chinle depositional system within this overall fluvial setting—evidence for the former presence of volcanic ash within the fluvial sand and mud. This material yields evidence for another interesting and important part of the Chinle story. Ash erupted from explosive volcanoes that

A three-toed fossil footprint of the dinosaur Dilophysaurus *in the Moenave Formation near Tuba City, Arizona*

existed in the mountains far to the south and west of the plateau. As this ash rained down on the Chinle river plains, it occasionally choked the streams with silica-rich debris and may actually have been the cause of the trees' ultimate demise. The dead tree trunks were then moved by floods into giant logjams that were buried by subsequent eruptions of volcanic ash and other sedimentary flood deposits. This specific sequence of events not only killed, transported, and concentrated the logs in silica-rich sediment, but quickly buried them so that they did not rot away. The ash eventually decomposed to clay, freeing the silica molecules in the ash, which then mineralized in the microscopic voids in the wood. This is how the logs ultimately became silicified (petrified with silica).

Today the weathered ash deposits form colorful claystones and are exposed in a frothy, popcorn-like surface of the Painted Desert. Each time the rocks get wet, the clay expands, forming this curious texture, which, incidentally, plays havoc with roads built upon it. Badlands have been carved into these fluvial and volcanic remains and form an awesome sight, referred to by some Native Americans as a "land of sleeping rainbows." Iron, manganese, and other metallic ions oxidized within the clay minerals give the rocks much of their vivid color.

Conditions became drier as the Late Triassic continued. Streams became more ephemeral, and local windblown dunes are preserved in the uppermost members of the formation. These latest Chinle redbeds were the harbingers of the arid conditions that would dominate the Colorado Plateau region during the Jurassic Period that followed.

Dunes and the Great Sand Pile: Jurassic

The three-part Glen Canyon Group exposed at its namesake along Lake Powell. The lower cliff is the Wingate Sandstone, the middle ledges are of the Kayenta Formation, and the upper cliff is composed of the Navajo Sandstone, all of Jurassic age.

Drought, absolute and relentless, dominated the Colorado Plateau region during the Early and Middle Jurassic. Three of the greatest eolian ergs of the ancient past were deposited during this time of drought. Widespread eolian conditions require several factors to develop: aridity that suppresses vegetation, an open landscape, a source for the sand, and lots of wind—steady and strong. The sand was supplied by the many rivers that drained the once-mighty Appalachian system to the east. These rivers flowed west and northwest and delivered sandy sediment to the region. Many of these rivers had intermittent or ephemeral flow as they dried up on the hot Jurassic plains and strong, consistent winds blasted the parched floodplains and blew the sand toward the southeast into huge dune fields. Occasional river floods then swept some of the dune sand back to the northwest. This pattern of alternating fluvial and eolian transport occurred many times during the Jurassic Period. The well-rounded sand grains and increased sorting of the sediment (tendency toward accumulation of grains of the same size) are both evidence of this.

The first of the Jurassic sand seas left a spectacular cross-bedded sandstone known as the Wingate Sandstone. (Although paleontological evidence suggests that some of the lower part of the Wingate may be Triassic, we will discuss the entire unit as being dominantly Jurassic.) This reddish-orange, cliff-forming sandstone predominates in many areas on the central Colorado Plateau, including the northern parts of Glen Canyon National Recreation Area, Capitol Reef and Canyonlands national parks, and much of the Navajo Indian Reservation. Upon this modern landscape, the Wingate cliffs form fabulous sheer walls that are oftentimes veneered with black or gunmetal-blue desert varnish. This varnish, formed in part by the metabolic process of bacteria on the rock, takes many thousands of years to develop and attests to the Wingate's durability and hardness. The resulting stately profile of the Wingate cliffs belies their loose, sandy origin on the desert plains and dune fields of Jurassic North America.

Wingate cliffs, Kayenta ledges, and Navajo domes, all of the Jurassic period at Tower Butte, Canyonlands National Park, Utah

As the Wingate dunes were busy burying the central plateau region, ephemeral streams existed to the southwest. These streams flowed northwest and deposited mud, coarse sand, and conglomerate, which today comprise the coeval Moenave Formation (Vermilion Cliffs and Grand Staircase–Escalante national monuments, and Zion National Park). Interfingering of the two deposits, with the eolian Wingate Sandstone to the northeast and the fluvial Moenave Formation to the southwest, occurs below the surface just east of the Echo Cliffs in Arizona. The Moenave stream system furnished much of the sand that fed the Wingate erg. This detailed interfingering of the deposits allows for a precise reconstruction of the Early Jurassic landscape, with rivers and dunes coexisting on different parts of the plateau surface.

As time passed, the Wingate dunes were overwhelmed everywhere by a sandy, braided fluvial system preserved as the Kayenta Formation. These Kayenta rivers buried the Wingate dunes with deposits of sand, mud, and intraformational conglomerate (with clasts derived from that very same deposit). Subsidence created the accommodation space that

The eolian dune field at Sossusvlei, Namibia, is an example of how the Colorado Plateau may have looked at times during the Jurassic Period.

put the Wingate Sandstone into storage beneath these river deposits. In addition to subsidence in the Kayenta fluvial basin, a coeval pulse of uplift in the Ancestral Rockies allowed for river sand to dominate the depositional system at this time. Uplift of the Ancestral Rockies may have also acted to create the slightly wetter climatic conditions that favored more fluvial conditions during Kayenta time. The Kayenta

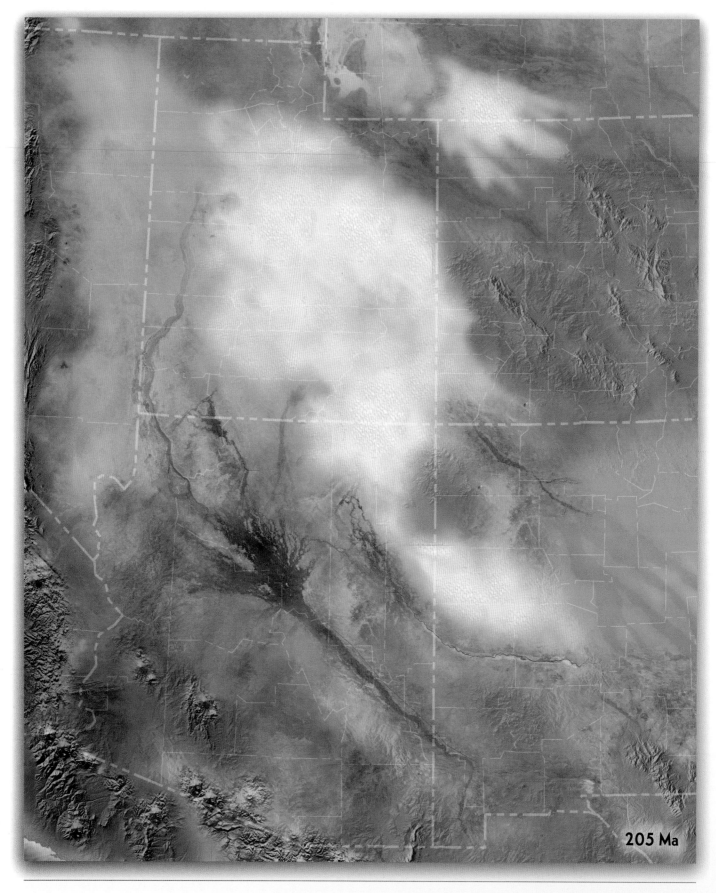

205 Ma

Above: Early Jurassic paleogeography (205 Ma) when the eolian Wingate Sandstone and the fluvial Moenave Formation (Dinosaur Canyon Member) were deposited. The lateral boundary between these two contrasting depositional systems fluctuated back and forth and preserve the intimate details for this earliest Jurassic time slice on the Colorado Plateau.

Opposite: The Early Jurassic landscape (200 Ma). Kayenta streams came from both the Ancestral Rockies and the Appalachians and flowed west across the Colorado Plateau. These rivers were perennial but flowed across arid country.

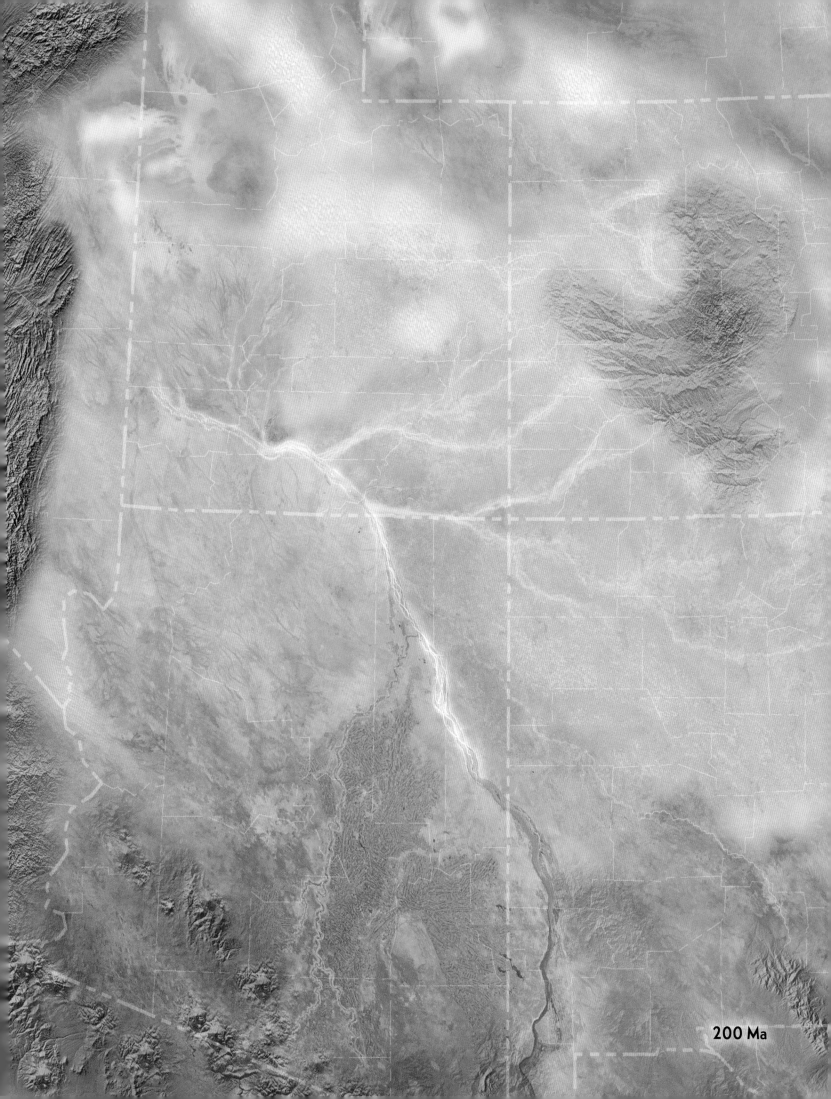

200 Ma

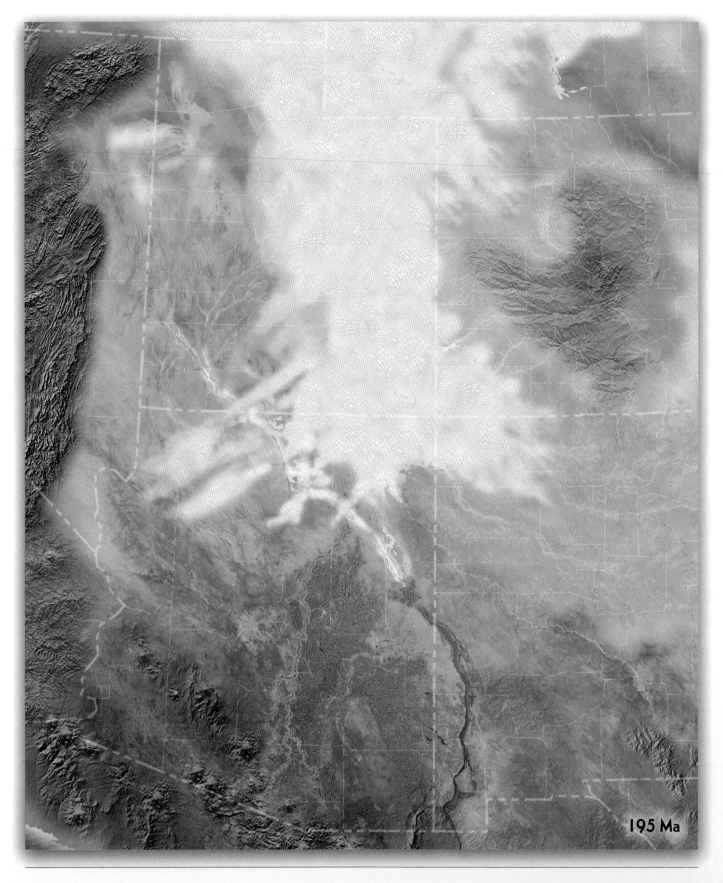

195 Ma

Above: Early Jurassic paleogeography at 195 Ma. Because the contact between the Kayenta and Navajo formations is gradational, a complex interplay between fluvial and eolian conditions is preserved. The interval is hundreds of feet thick and suggests that the "battle" between these two depositional systems was relatively long lived.

Opposite: Early Jurassic landscape at 190 Ma when the great dune field of the Navajo erg spread across the Colorado Plateau. Note that the dunes migrated far to the south (southern Arizona), into the volcanic arc. The Navajo Sandstone represents perhaps the largest and most voluminous ancient eolian sand deposit on Earth.

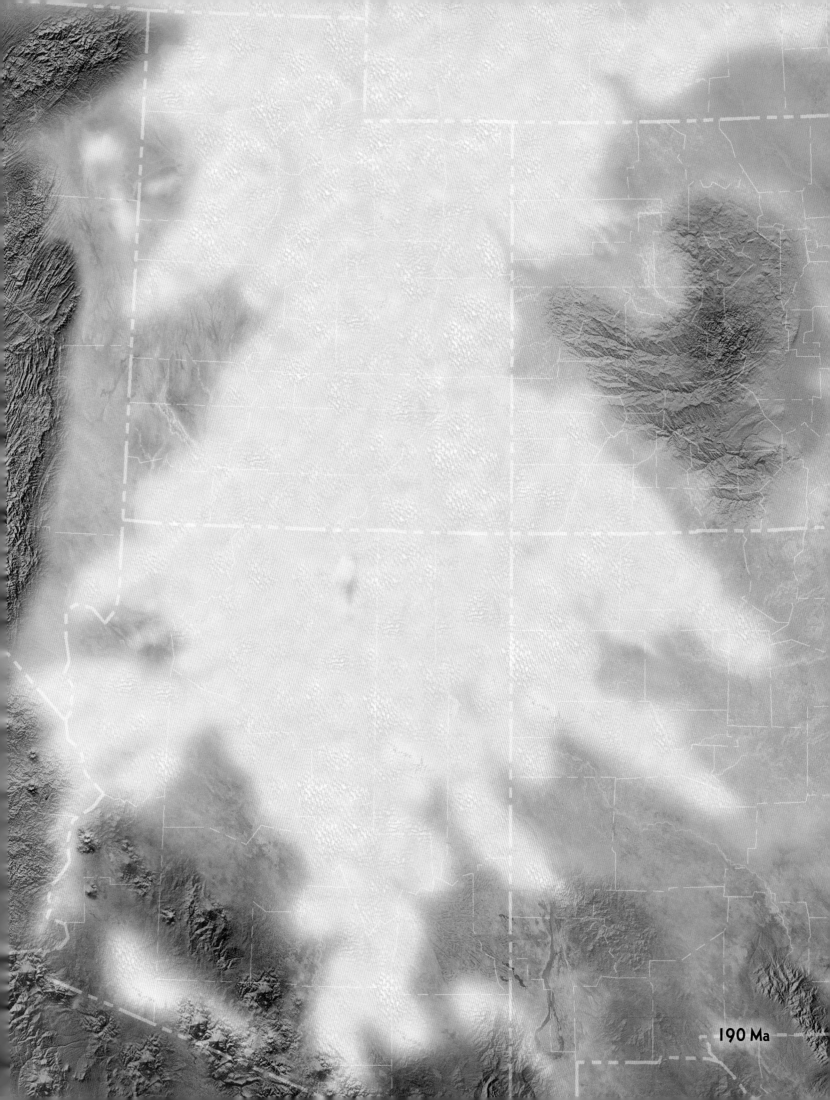

190 Ma

Formation forms easily recognized but often overlooked ledges and slopes that are sandwiched between two prominent cliff formers above and below it (Navajo and Wingate sandstones respectively). All three formations are collectively known as the Glen Canyon Group. The Kayenta is exposed beautifully at Canyonlands, Zion, and Capitol Reef national parks, and at Navajo and Colorado national monuments.

The Navajo Sandstone exposed along the Colorado River in Glen Canyon near Page, Arizona

Gradually, dune conditions reappeared on the landscape, advancing across the broad Kayenta floodplain. The birth of the earth's greatest fossil eolian desert, preserved as the Navajo Sandstone, was underway. This Lower Jurassic sandstone covers much of the Colorado Plateau region and equivalent deposits spilled southward and westward into southern Arizona, Nevada, and California, where they are called the Aztec Sandstone. In Wyoming, coeval deposits are called the Nugget Sandstone. Before being eroded later in the Jurassic, the Navajo probably also covered large parts of Colorado and New Mexico. It is truly one of the great eolian deposits in all of geologic time. Fossils, though sparse, are present in the Navajo and include rare dinosaur trackways and even rarer bone material. This was the time of the huge sauropods, and certainly a few of them wandered the fringes of the Navajo desert erg. More abundant however, are trace fossils left by crawling and burrowing organisms, probably worms and invertebrate larvae. Casts of trees have recently been described from exposures near Arches and Canyonlands national parks.

An example of a limestone deposit within the Navajo Sandstone near Moab, Utah. The limestone originated when the Jurassic water table rose above the surface of the dune field and created an ancient oasis.

As the large sandy Navajo dunes were built across all of Utah and northern Arizona, a few Kayenta streams held on and continued to flow across north-central Arizona. These streams encountered scorching conditions on the desert plains of southwestern Utah and as the water evaporated, it left dry river channels exposed to the hot winds. These winds carried the loose, dry sand toward the east, and a second episode of eolian-fluvial cycles occurred during the Early Jurassic. Eventually, a great thickness of this sediment accumulated as the southwestern part of the Colorado Plateau rapidly subsided. More than 2,500 feet (762 m) of eolian sand comprises the Navajo Sandstone in Zion National Park.

The dominantly sandy Navajo also contains thin beds of gray limestone, not uncommon in some outcrop areas. These may seem out of place within such a predominantly desert setting but can be attributed to the presence of ancient oases in the desert. The limestones are never more than three feet (1 m) thick, are thinly bedded, and show crinkly texture within the beds—an indication that they were formed by algae. The interpretation is that the algae formed in ponds in the interdune environment, those areas located between the towering dunes, which were flat and oftentimes flooded by high groundwater tables. Algae gained a foothold in these shallow ponds and secreted the limestone. The evidence for these interdune ponds could explain the tree casts and occasional sauropod footprints within the otherwise uninhabitable Navajo desert.

The top of the Navajo Sandstone is truncated by a major regional unconformity, which signals the end of Early Jurassic deposition and separates rocks of the Lower and Middle Jurassic. How much Navajo sand was removed during this erosion episode will never be known but it probably was substantial, especially in the eastern portions of the Colorado Plateau.

The eolian Temple Cap Sandstone lies on top of the Navajo Sandstone but is restricted entirely to the Zion National Park area, where it caps the towers and temples of the Virgin River (and thus the derivation of its name). The restricted presence of the Temple Cap suggests this was an area of rapidly subsiding terrain just prior to the onset of the unconformity.

The Middle Jurassic was a time of complex sedimentation and began with deposition of the Page Sandstone in north-central Arizona, in the Lake Powell area. The Page Sandstone originated in eolian dune settings identical to those of the Navajo Sandstone. The two formations are virtually indistinguishable, but an inconspicuous chert-pebble conglomerate marks a subtle unconformity. At this time, sea level began to rise again, covering only the western half of the plateau region. The long, narrow Sundance Sea (named after a limestone formation of that name in northern Utah) extended down from the north in shallow, saline environments. The Carmel Formation, a heterogeneous mix of red sandstone, mudstone, gypsum, and limestone, formed near the fluctuating southern shoreline of the Sundance Sea on the southern Colorado Plateau region. At times, the seaway expanded to the southeast and infringed into the Page eolian deposits. The dunes gradually expanded when the sea withdrew to the northwest. A complex stratigraphy of interbedded eolian sandstone and marine deposits thus forms much of the lower Carmel Formation (Grand Staircase–Escalante National Monument and Glen Canyon National Recreation Area, along U.S. Highway 89 between Page, Arizona, and Kanab, Utah).

Volcanic activity raged at this time beyond the southwest margin of the Colorado Plateau from near present-day Yuma, Arizona, to Las Vegas, Nevada. Subduction of oceanic crust beneath the southwestern edge of North America created a long and active volcanic arc that extended the entire length of North and South America. Termed the Cordilleran Arc, this feature is still active in many parts of the Americas today, notably in the Andes, Central America, and Cascade Mountain volcanic regions. Volcanic ash from this arc was spread widely across the Colorado Plateau region in the Middle Jurassic and is preserved within beds of the Page and

Large-scale cross-beds in the Navajo Sandstone in Zion National Park. Cross-beds dip right to left indicating the paleo-wind direction during the Jurassic.

An example of large-scale cross-bedding from the Middle Jurassic Page Sandstone in north-central Arizona

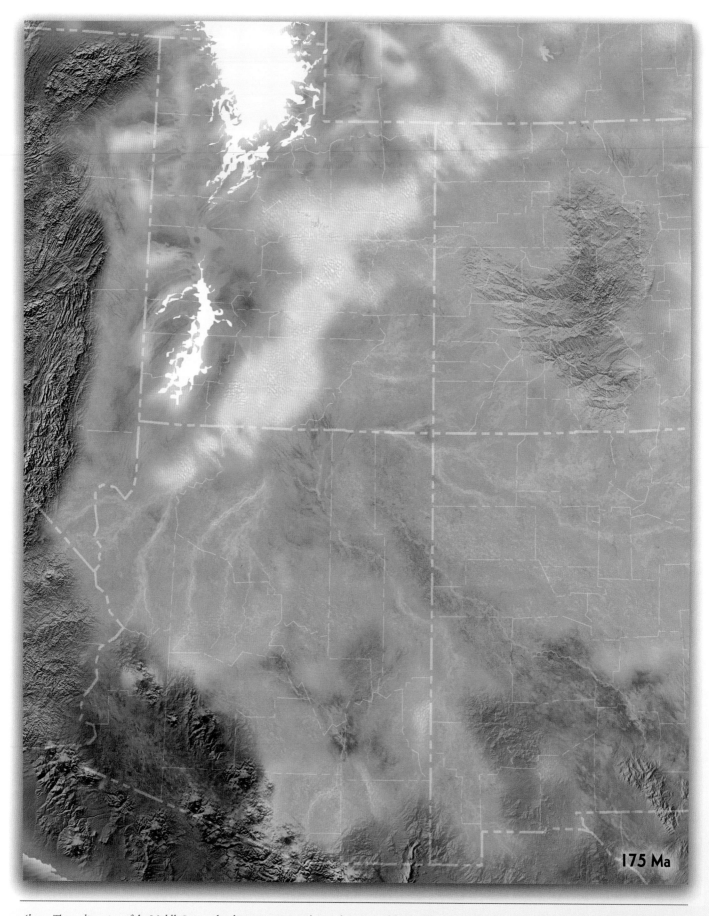

175 Ma

Above: The earliest view of the Middle Jurassic landscape at 175 Ma during deposition of the Temple Cap Sandstone, preserved today only near Utah's Zion National Park on the Colorado Plateau.

Opposite: The Middle Jurassic paleogeography at 170 Ma marks a time of significant change on the landscape. Several times a narrow, restricted seaway entered the Colorado Plateau region from the north and deposited the marine Carmel Formation and the coeval eolian Page Sandstone.

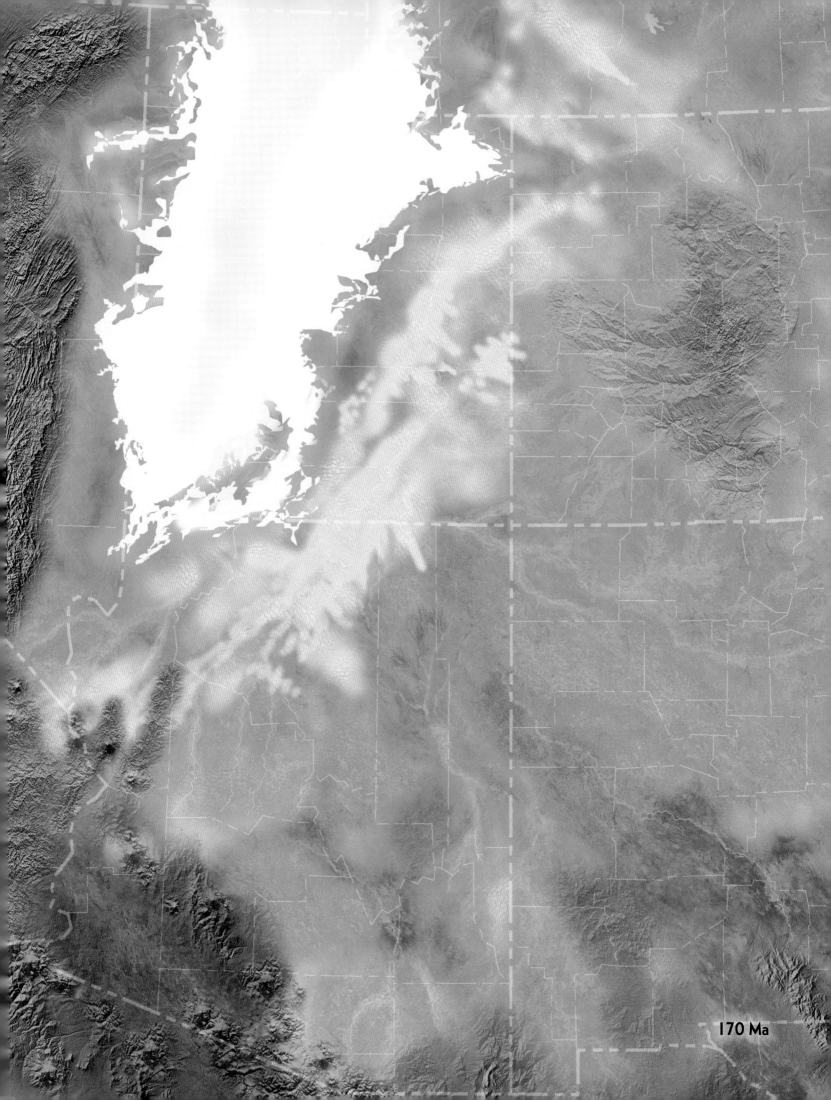

170 Ma

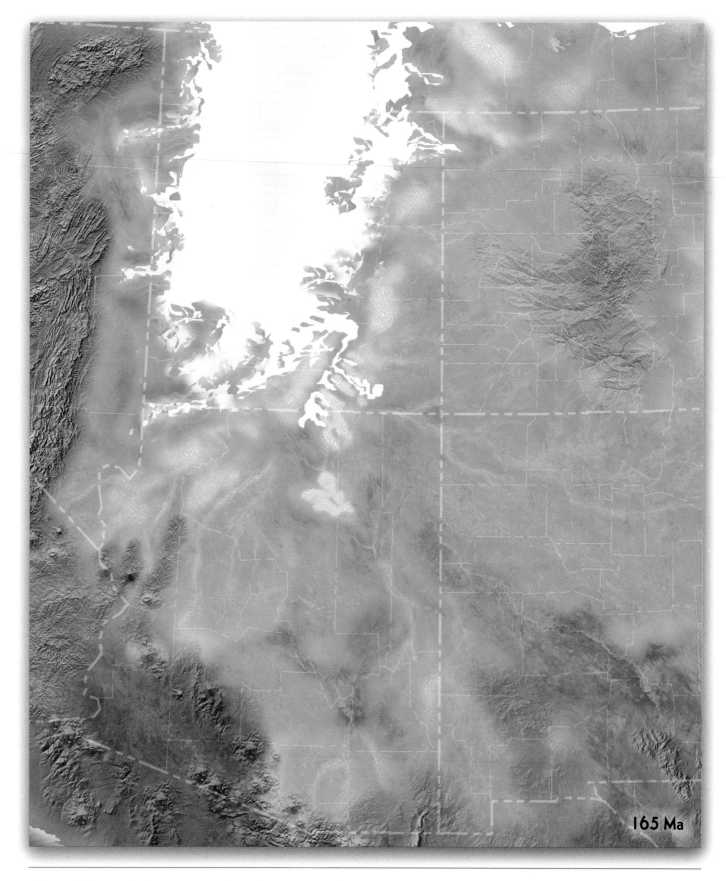

165 Ma

Above: Middle Jurassic landscape (165 Ma). The Sundance Sea at one of its greatest extents deposited the Upper Carmel Formation on the Colorado Plateau. Rivers from the Cordilleran Arc (left) carried volcanic grains to the coastal plain in eastern Nevada, while dune fields south of the sea migrated across arid mudflats. Evaporites were deposited in restricted coastal flats in central Utah.

Opposite: The Middle Jurassic at 160 Ma, when the Entrada dune field marches across the Colorado Plateau region. Deposits of Entrada Sandstone are different across a line drawn from southwest to northeast Utah—eolian dunes and sabkhas were southeast; coastal sabkhas and restricted muddy, hypersaline seas were northwest.

160 Ma

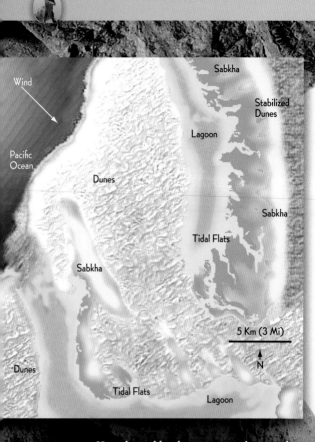

Hypothetical landscape setting showing the relationship between dunes, sabkhas, and shallow marine environments as preserved in the San Rafael Group of the Colorado Plateau. This map is based on the modern setting at Guerro Negro, Baja California, Mexico.

Middle Jurassic ammonites at the University of Tübingen, Germany. Although museum specimens like these are extremely rare, and nothing like these has been found in the Colorado Plateau, these same ammonites swam in the Sundance Sea of Utah and Wyoming.

Carmel deposits. These volcanic horizons can be easily dated by radiometric methods and thus make good time markers in otherwise undatable rocks. Therefore, even though the Page Sandstone may be barren of fossils, it is easily dated and correlated with the fossiliferous Carmel Formation, which is locally abundant in mollusks and crinoids.

In the latter part of the Middle Jurassic, eolian dunes once again returned to the region depositing the Entrada Sandstone. This sand sea was almost as grand as the Navajo erg that preceded it by several tens of millions of years. Although the Entrada was deposited in an eolian setting, sabkha deposits are common in the unit and relatively high water tables created these wet ground conditions. One of these sabkha deposits is present along the Utah-Colorado state line and is known as the Wanakah Formation. At one time the Entrada Sandstone covered most of Utah, southern Wyoming, Colorado, northern Arizona, and New Mexico. The arches in Arches National Park, as well as the goblins in Goblin Valley State Park, are carved from this sandstone. At Black Canyon of the Gunnison National Park in western Colorado, the Entrada Sandstone sits on top of Precambrian crystalline rocks, documenting the specific time when the Ancestral Rockies were completely beveled and buried—some 165 million years ago and almost 145 million years after they made their initial appearance on the landscape.

In latest Middle Jurassic time, a restricted seaway returned once again to the region. Marine sandstone and mudstone of the Curtis Formation reached as far south as the Waterpocket Fold area in Capitol Reef National Park. The Curtis locally contains fossils of clams and cephalopods. To the south, the coeval Romana Sandstone and Summerville Formation accumulated in tidal, sabkha, and eolian settings on this coastal plain. These rocks are barren of fossils, suggesting arid conditions. They are exposed in southern Utah in Grand Staircase–Escalante National Monument and Glen Canyon National Recreation Area.

One last Middle Jurassic period of erosion and nondeposition followed the Curtis–Romana–Summerville interval. The resulting unconformity may be the result of tectonic events occurring far from the plateau region to the southwest. Evidence comes from a pronounced change in paleocurrent (direction of flow) data that has been noted between Middle and Late Jurassic stream deposits. Recall that from Early Triassic through Middle Jurassic time (some 70 million years) streams flowed generally from southeast to northwest. These streams drained the Ancestral Rockies, the Appalachians, and highlands in southern Arizona, the last of which belong to the upper member of the Carmel Formation. Beginning in the latest Middle Jurassic and continuing through the Late

Jurassic, streams began to drain from the southwest toward the northeast in the early phases of the newly formed Cordilleran Arc. By the Late Jurassic, paleocurrents show streamflow toward the northeast from Nevada and western Arizona, areas that had been uplifted in the Cordilleran Arc, becoming the source area for Late Jurassic streams. A broad look at the variable directions of river flow through time shows a significant shift in source area from the Early Triassic to the Late Jurassic.

A spectacular specimen of the fossil Stegosaurus, *a typical dinosaur from the Morrison Formation*

Late Jurassic streams deposited the famous Morrison Formation, one of the world's greatest dinosaur graveyards. The streams drained uplands in Nevada and central Arizona that were the result of the Nevadan Orogeny (a mountain-building event that was part of the Cordilleran Arc). Radiometric data from Nevada and California date the orogeny at approximately 150 million years ago. The Nevadan Orogeny resulted from crustal stresses caused by the subduction of oceanic plates in the Pacific Ocean that collided with western North America. The uplift in Nevada exposed Paleozoic rock units to erosion, fragments of which were transported by Morrison streams across the Colorado Plateau region. Some of these rock fragments, especially some chert grains, can be tied directly to the specific Paleozoic units that were being eroded in Nevada and California.

Environmental conditions during Morrison time were varied and complex, leaving a similar set of perplexing deposits. The rocks are dominated by fluvial sandstone, conglomerate, and mudstones. Like the Chinle Formation before them, volcanic ash was erupted from the highlands in Nevada and spread across the fluvial floodplain. It has since weathered into claystone leaving badlands in many areas of the plateau (Capitol Reef National Park and Dinosaur National Monument). Stream systems were likely humid close to the mountain front but may have become more arid as they coursed downstream to the northeast. Evidence for this comes from units such as the Bluff Sandstone, exposed along the San Juan River near Bluff, Utah. It is an eolian sand deposit that most likely was situated within the braided fluvial system of the Morrison Formation. Details such as this provide unequaled glimpses into Late Jurassic geography. At one time the Morrison Formation probably covered most of the Colorado Plateau; indeed it extended over the future site of the Rocky Mountains and well onto the Great Plains.

Morrison Formation in Capitol Reef National Park, Utah. Sandstones of the Salt Wash Member are visible in the foreground, as are colorful claystones of the Brushy Basin Member in the middle distance. These deposits likely originated as river sands and muds, as well as weathered volcanic ash.

The Morrison Formation is world famous for its dinosaur fossils. Hundreds of partial and several complete skeletons have been recovered, especially from

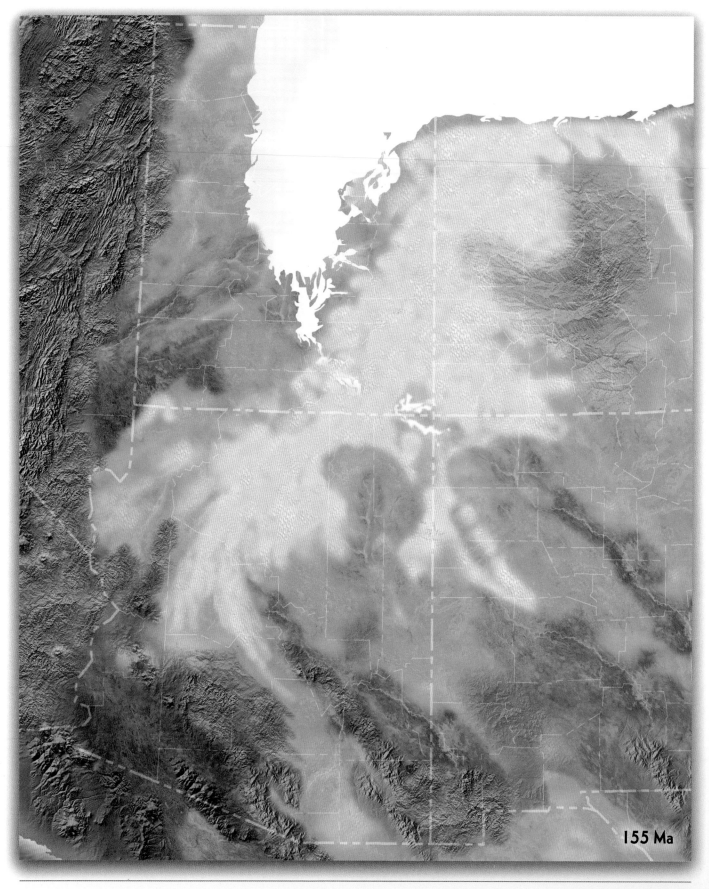

155 Ma

Above: Middle Jurassic paleogeography (155 Ma). The final Jurassic sea is recorded in the Curtis Formation in central Utah and the onshore sabkha deposits of the Summerville Formation to the east and south. Initial gravel deposits from the Nevadan Orogeny to the west are in marked contrast with those of the Kayenta Formation, which came from the east.

Opposite: Late Jurassic landscape elements (152 Ma) of the Morrison Formation. A retreating seaway allowed red sandstone and mudstone of the Tidwell Member to be deposited in coastal plain, lacustrine, and eolian environments.

152 Ma

150 Ma

Above: The Late Jurassic at 150 Ma. Rivers left sediments belonging to the Salt Wash Member of the Morrison Formation on the central Colorado Plateau. This unit is one of the planet's greatest dinosaur graveyards—a true Jurassic Park! To the southeast near the Four Corners, eolian conditions existed and deposited the Bluff, Junction Creek, and "Zuni" sandstones.

Opposite: The Late Jurassic at 145 Ma, when the Brushy Basin Member was deposited. As the Jurassic Period ended, the climate became more humid and fluvial conditions persisted. A large alkaline lake existed in the eastern portion of the region.

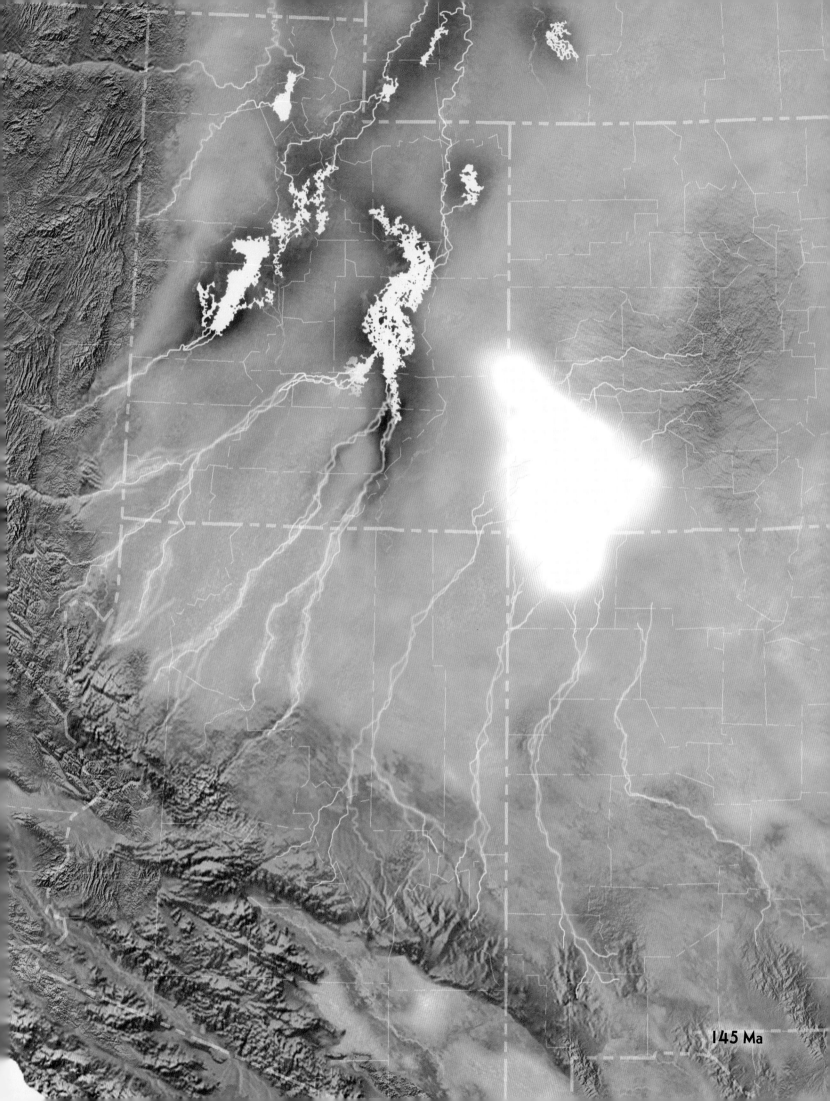

145 Ma

Entrada Sandstone at Delicate Arch, the symbol of Arches National Park, Utah

Dinosaur National Monument in northeast Utah and northwest Colorado, and the Cleveland-Lloyd Quarry near the San Rafael Swell in Utah. For many years, the rich dinosaur beds at Dinosaur National Monument were an enigma to geologists—the bones were articulated and many different species were present. This is an odd occurrence, and an early interpretation was that a catastrophic flood had carried the corpses along on rushing floodwaters, even though this would seemingly have caused the disarticulation of the bones. Recently, it was determined that a host of individual animals had probably met their demise while hunkering down next to an evaporating water hole that eventually dried up in the hot Jurassic sun. The animals may have been drawn to this water hole for survival, only to die a slow death of thirst next to the parched pond. Such is the detail of preservation that can be found in these deposits. Many other kinds of fossils are also found within the Morrison, including plants, small mammals, and numerous invertebrates.

Location of Triassic and Jurassic Rocks

Triassic rocks are perhaps the most widely exposed rocks on the Colorado Plateau. In Arizona they form a broad outcrop band along the valley of the Little Colorado River from St. Johns to Cameron, then northward along the Echo Cliffs to Lees Ferry. This is the land of the Painted Desert, so named for the brilliant Triassic rocks of the Moenkopi and Chinle formations. These rocks are found in Petrified Forest National Park and are what cap the low mesas and buttes between Winslow and Holbrook, and form the intricate and colorful hills and "tepees" along U.S. Highway 89 southwest of Tuba City. They crop out in the Flagstaff area where they have been quarried to furnish stone from the Moenkopi Formation for buildings as far away as the Brown Palace Hotel in Denver and an earlier incarnation of the Los Angeles City Hall. They form the base of the Echo Cliffs and the associated badlands where the swelling clays in the Chinle Formation are responsible for the bumpy sections of U.S. Highway 89 north of Cameron. They underlie Chinle Valley in northeast Arizona, from which the formation received its name. In New Mexico, Triassic rocks follow I-40 from Gallup to Grants at the base of the Wingate cliffs. In Colorado, Triassic rocks crop out in the Dolores River valley and the salt anticline region. In Utah, Triassic rocks ring the San Rafael Swell, the Teasdale uplift, and the Circle Cliffs where they form intricate badland country below imposing Jurassic cliffs. They are widely exposed in the Canyonlands region and southward across the Monument upwarp and Comb Ridge to Monument Valley, where they cap the very tops of the "monuments." Along the Arizona-Utah border, Triassic rocks are exposed in the deeper parts of Glen Canyon (although many Triassic outcrops lie beneath Lake Powell), and they follow the base of the Vermilion Cliffs from Lees Ferry to Zion National Park. Triassic rocks form the spectacular badlands at Old Paria and the Chocolate and Shinarump cliffs near Kanab, Utah. Below Zion Canyon, they follow the Virgin River valley to the edge of the Colorado Plateau and comprise the banded mesas and badlands including Hurricane Mesa. In Dinosaur National Monument, they form the low, red valleys around Split Mountain; at Flaming Gorge National Recreation Area they provide some of the color for which the gorge was named.

Jurassic rocks form many cliffs and canyons on the plateau and are exposed about as extensively, and follow the same trends as Triassic rocks described in the previous paragraph. In northern Arizona and southern Utah, they cap the Vermilion, White, Echo, Circle, and Orange cliffs. They form the tops of the Island in the Sky and Dead Horse Point in the Canyonlands region. Jurassic rocks compose the jagged cliffs that rim the uplifts and form the backbones of the monoclines: the Grand Staircase–Escalante's Cockscomb (on the East Kaibab upwarp), Comb Ridge (the Monument upwarp), San Rafael Reef, Capitol Reef, and Waterpocket Fold. They form the walls and slot canyons of Glen Canyon and the cliffs around Moab and the adjacent Colorado River; they form the sheer cliffs at Colorado National Monument and the cliffs that rim the salt anticlines in western Colorado and adjacent Utah. Jurassic rocks form the mesas west of Monument Valley and the canyons in Navajo National Monument. The slickrock country around Escalante and Moab is mainly carved in Jurassic rocks. They also crop out in Dinosaur National Monument. There, as well as at Cleveland-Lloyd Quarry near the San Rafael Swell, they contain tremendous numbers of dinosaur skeletons. Jurassic rocks are also spectacular arch and natural bridge formers: Rainbow Bridge (the world's largest); Hickman Bridge (Capitol Reef); the arches of White Mesa, Arizona; Arches National Park; the arches in the Escalante area; and many others. Last, but hardly least, Jurassic rocks comprise some of the greatest cliff faces in the world, the towering 3,000-foot-high (900m) walls of Zion Canyon.

Middle Jurassic rocks near Lake Powell on the Arizona-Utah border

Chapter Summary

The fragmentation of Pangaea during the Triassic initiated significant changes on the Colorado Plateau. As North America began its westward drift into oceanic plates, the area became slightly elevated above sea level and continental deposits dominated for over 100 million years thereafter. Dry, scorching trade winds caused widespread deposition of eolian and ephemeral fluvial deposits, within some of the most colorful rock formations anywhere. Three major sand seas (the Wingate, Navajo, and Entrada) were present during Middle Jurassic time. By the Late Jurassic, the Nevadan Orogeny had raised mountains to the west, and these east-directed fluvial systems of the Morrison Formation closed out the Jurassic with dinosaur-rich sediments. This set the stage for the more humid conditions that would characterize the Cretaceous.

Fossil of an Ichthyosaur, a common marine reptile that swam the Mesozoic seas of the Colorado Plateau

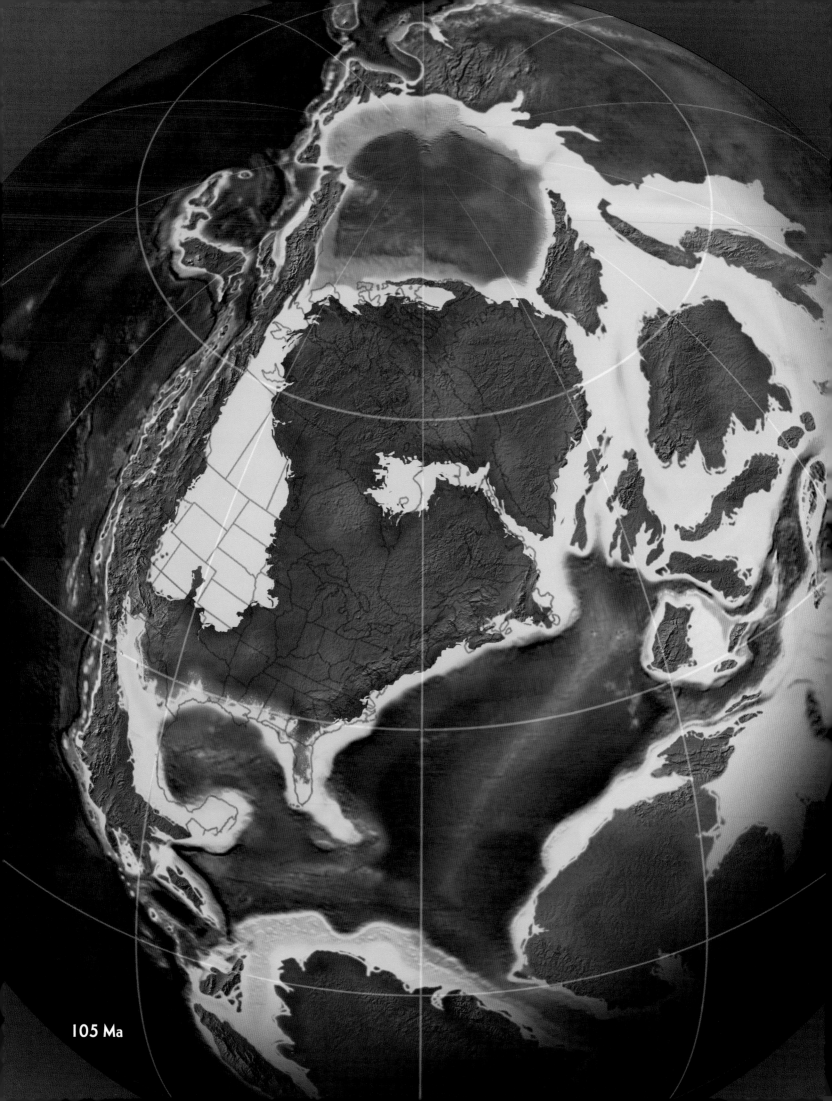

105 Ma

Chapter Six
Mountains and Continental Seaways
Cretaceous: 145 to 65 Million Years Ago

The initial breakup of Pangaea into smaller continental fragments began during the Triassic, but many of the extreme consequences of this event were not felt until the Cretaceous. While Pangaea existed, heat trapped beneath the supercontinent kept it relatively high in comparison to ocean levels. But as the continents became more widely separated, this built-up heat escaped, causing the continents to gradually become lower relative to sea level. This resulted in a series of great transgressions across all of Earth's continents during the Cretaceous. As the Atlantic Ocean expanded, North America plowed its way westward into parts of the Pacific Ocean crust. The Cordilleran Arc and subduction zone strongly influenced the geology of the western part of the continent—in fact, its influence continues to this day as the Juan de Fuca plate subducts beneath Cascadia from northern California to British Columbia, and the Cocos plate dives beneath Mexico and Central America.

At the beginning of the Cretaceous great changes began to affect the Colorado Plateau region. One of these changes was a tremendous uplift to the south and west in southern Arizona, California, Nevada, and western Utah. This uplift is recorded on the plateau by a regional unconformity between Jurassic and Cretaceous strata, which progressively bevels (planes off by erosion) older strata toward the southwest. Previously deposited strata were tilted up on the southwestern side of the Colorado Plateau. For example, on the Mogollon Rim, the Dakota Sandstone sits atop the Kaibab Formation. A progressively beveled sequence is found from Zion eastward toward Lake Powell. At Lake Powell, Cretaceous rocks sit on the Morrison Formation, but just a couple of miles west they rest on the underlying Romana Sandstone. Thirty miles (48 km) farther west and near the Paria River (Grand Staircase–Escalante National Monument), Cretaceous rocks rest on the still older Entrada Sandstone. And another twenty-five miles (40 km) to the west, the Entrada is completely eroded and Cretaceous rocks rest on the Upper Carmel Formation. Finally, in Zion National Park, at the very western edge of the Colorado Plateau, Cretaceous rocks overlie the middle part of the Carmel. These relationships document how the strata were uplifted to the west at the beginning of the Cretaceous.

90 Ma

Opposite: Global view of North America during the Cretaceous Period, 105 Ma. The continent has broken free from South America and Africa. The West Coast is tectonically active and mountainous, and a huge seaway begins to inundate the continent from the north.

Right: Late Cretaceous (90 Ma) globe showing one of the greatest continental floodings in geologic history. North America eventually became bisected into two landmasses during this major flooding.

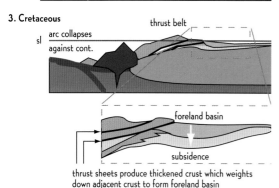

Stages in the formation of the Cretaceous foreland basin, western North America: 1. Early and middle Paleozoic passive margin—great thicknesses of sedimentary rocks accumulate; 2. Change to active margin with bordering island arcs; 3. Arcs and accreted terranes collide and collapse against North America during the Sevier Orogeny. Inset shows how thickened thrust sheets (Paleozoic rocks) depress the crust, causing subsidence of foreland basins.

Aerial photo of the Cretaceous rocks in the Book Cliffs near Green River, Utah. This section matches the profile of the diagram opposite bottom.

In conjunction with the Late Jurassic–Early Cretaceous uplift in areas to the south and west, the plateau region began to rapidly subside. As the older Nevadan Orogeny became quiescent, new and different kinds of stresses built up in the crust along the western margin of North America. These stresses created a large thrust belt, whereby large sections of solid rock were pushed horizontally over adjacent rocks to the east. These huge thrust faults, typically putting older rocks on top of younger rocks, occurred deep below the surface; however, subsequent erosion has exposed some sections of these faults on the earth's surface today. This mountain-building event, called the Sevier Orogeny, rapidly thickened the western crust as slices of rock were horizontally shoved and vertically stacked one on top of the other. The thrust belt lies west of the Wasatch line (the plateau edge) and stretches from eastern Idaho and northern Wyoming to southern Nevada and southeastern California.

Colorado Plateau strata were thus not directly involved in the thrusting. However, the weight of the thickened crust in eastern Nevada and western Utah exerted a downward force on adjacent lands to the east. This resulted in subsidence of the plateau region in a type of tectonic basin known as a foreland basin. These are basins that sit in the foreground of the uplift when viewed from the continent. In conjunction with the development of this foreland basin, one of the highest sea-level episodes in the geologic record began. The cause of this rise is intimately related to increased volcanic activity in the western Pacific Ocean. These eruptions caused Pacific Ocean crust to swell, and seawater was displaced from the normally deep ocean basin onto the edges of all of the earth's continents. High sea level, coupled with rapid subsidence in the foreland basin, produced a great marine flooding (although it was gradual by human standards). Also during the Cretaceous, the regional climate became much more humid than it had been during the previous 150 million years. The Cretaceous signaled an end to the dominantly arid climatic conditions that prevailed for most of the Mesozoic prior to this time. Although reasons for this change were complex, a major factor was the northward drift of North America, which by Late Jurassic time had entered the midlatitude belt of westerly winds and wetter climates.

In contrast to the bright reds and oranges of the plateau's Triassic and Jurassic rocks, Cretaceous rocks are drab-colored tan, gray, or green, reflecting an increase in the organic content of the rocks, which itself is a result of the increased vegetation and humidity of the time. In spite of their subdued colors, these rocks formed under very dynamic conditions—the repeated pulses of mountain uplift (which caused variations in sedimentation rates) and rapid sea level changes.

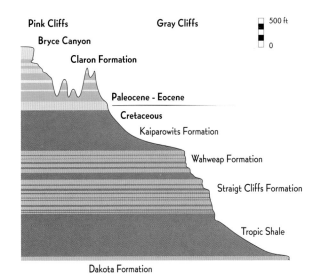

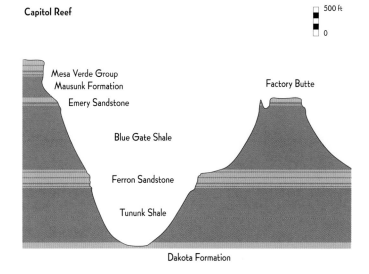

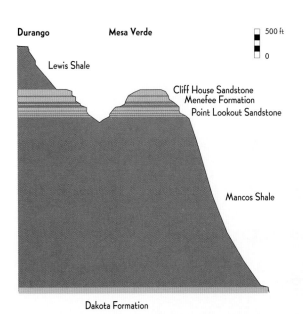

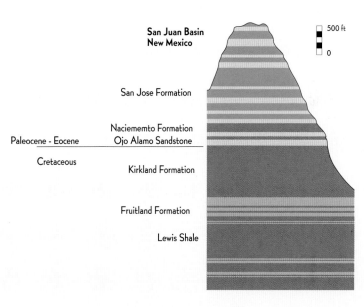

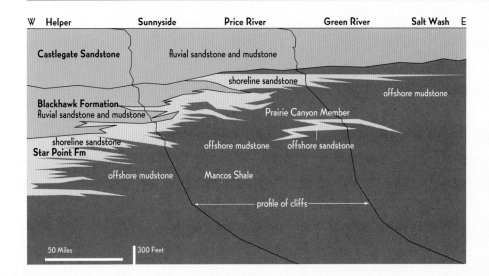

Above: Columnar sections of Cretaceous and Cenozoic rocks at four locations on the Colorado Plateau

Left: An example of the Cretaceous depositional patterns at the Book Cliffs in Utah. Marine mudstone is gray, offshore sandstone peach, shoreline sandstone yellow, and fluvial sandstone tan. A photograph of the profile at Green River is shown on the opposite page.

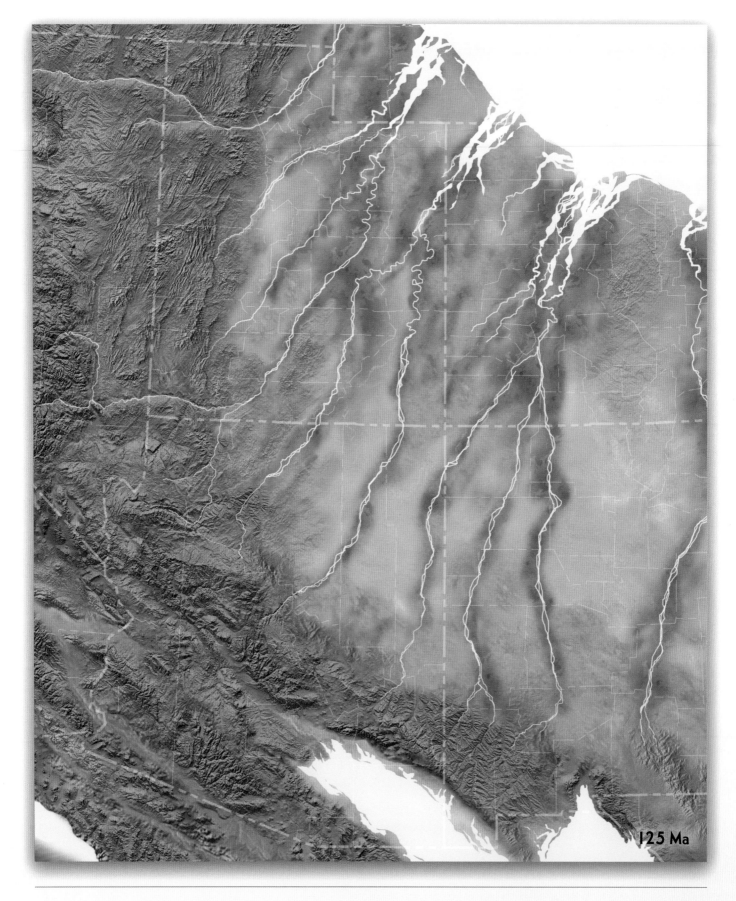

125 Ma

Above: Early Cretaceous (Barremian) paleogeography (125 Ma). (Detailed subdivisions of the Cretaceous are terms originating in Europe but used globally.) The southern Colorado Plateau was undergoing erosion as fluvial gravel was being deposited in central Utah as the Cedar Mountain Formation.

Opposite: Early Cretaceous (Albian) paleogeography (105 Ma). As the Sevier Orogeny matured, the Colorado Plateau region subsided and the Cretaceous Interior Seaway entered from the northeast. Deltas, shorelines, bars, and swamps left deposits known as the Dakota Sandstone. The peninsula in central Colorado was eventually drowned by the rising sea.

105 Ma

A scene from the Amazon River in South America is reminiscent of some of the Cretaceous landscapes on the Colorado Plateau.

A coal seam in the Dakota Sandstone at Coal Mine Canyon in northeastern Arizona formed in thickly vegetated swamps similar to that shown in the Amazon photograph above.

Mountains, Coal Swamps, and Epeiric Seas: Cretaceous

Cretaceous paleogeography was extremely varied. High mountains covered all of Nevada and western Utah and a broad seaway flooded the Western Interior from central Utah to Illinois. This seaway bisected North America, connecting the Arctic Ocean with the Gulf of Mexico. The seaway is known as the Cretaceous Interior Seaway.

Parallel sets of varied environments existed between the mountains and the seaway. A broad eastward-sloping alluvial and coastal plain covered the future Colorado Plateau. Near the distal end of the coastal plain and just landward of the beach environment, there were coastal swamps thick with tropical vegetation. Offshore from the beaches were barrier bars located in the shallower parts of the seaway. Each of these environments ran parallel to the mountain front and shifted repeatedly in lockstep with each other. When sea levels rose, the coastal plain flooded and shifted all environments to the west. When sea level fell, the coastal plain was greatly expanded to the east as were the swamps, shoreline, and barrier bars in front of it. The rate of sediment influx also influenced the location of each environment. During periods of uplift more sediment was eroded, creating gravelly river and delta lobes that pushed all environments to the east. Layers of peat formed in coastal swamps and, with burial and compaction over time, the peat became coal. Some of the largest coal fields in the world occur in Cretaceous rocks of the Western Interior.

The Cretaceous Period is divided into Early and Late epochs. During the Early Cretaceous, subsidence and increased sediment influx (rather than high sea levels) were the main controls on deposition. However, there are no Lower Cretaceous rocks on the southern part of the Colorado Plateau, suggesting that this area was relatively elevated at this time. Lower Cretaceous rocks are found in central and northeast Utah, documenting more subsidence for this time period in that area. In Utah, the Lower Cretaceous Cedar Mountain Formation (Capitol Reef National Park and Dinosaur National Monument) is composed of fluvial sandstone and mudstone. A scattered and locally diverse fossil record is found in these Lower Cretaceous rocks; dinosaurs, early mammals, turtles, fish, invertebrates, and plants are known from these rocks. The fossils range from terrestrial to marine depending on local environmental conditions.

At the end of the Early Cretaceous, the Cretaceous Interior Seaway began its tremendous transgression to the south and west. The seaway began as a narrow body of water in the Great Plains region but as sea level rose it flooded areas farther west toward the future Colorado Plateau. The first beach and coastal plain deposits of this sea are called the Dakota Sandstone. As the name implies, this unit is quite widespread in the

Western Interior and was named for exposures in Dakota County, Nebraska. Its first incursion onto the plateau was in northeast Utah (Dinosaur National Monument) during the Early Cretaceous. The gradually rising sea eventually encroached onto the plateau proper at the start of the Late Cretaceous (100 million years ago), depositing a broad swath of shoreline sediment. The Dakota is time-transgressive; that is, it gets progressively younger to the south and west. Its age is Early Cretaceous in northeast Utah but becomes Late Cretaceous in the Grand Staircase–Escalante National Monument to the west. The Dakota Sandstone preserves coarse beach and fluvial sand, coal deposits, and shallow marine muds. It is famous for its oyster-shell reefs, the remains of which can often be seen sparkling in the late afternoon sun.

Above: Cretaceous shales and sandstones compose these cliffs at Mesa Verde National Park, Colorado. Capping the mesa is the type section of the Point Lookout Sandstone of the Mesa Verde Group.

As sea levels continued to rise, the Dakota beach sands transgressed farther southwest across the coastal plain. Slightly deeper water deposits followed in its wake and buried the Dakota. The Mancos Shale and its related rocks, Tropic Shale in southwestern Utah and Pierre Shale on the Great Plains, are more than a mile (1.6 km) thick in some places. Marine conditions covered almost two-thirds of North America and similar proportions of other continents at this time. The drab, gray shale forms odd, moonlike badlands across much of the Colorado Plateau (Glen Canyon National Recreation Area). Paleontologists at the Museum of Northern Arizona have recovered fantastic reptiles from these shales including giant swimming plesiosaurs, flying pterosaurs with fourteen-foot wingspans, and a unique land-dwelling dinosaur, called the therizinosaur, which most likely died on land and subsequently floated to a watery grave in the Tropic Shale. A diverse invertebrate fauna includes clams, ammonites, snails, bryozoans, and worms.

Below: Ammonites once flourished in the Cretaceous Interior Seaway, fossils of which can be found in rocks like the Mancos Shale seen in the slopes of the above photograph.

92 Ma

Above: Late Cretaceous (Turonian) paleogeography (92 Ma) during maximum transgression of Tropic Shale

Opposite: Late Cretaceous (Turonian) paleogeography (90 Ma) during regression of Ferron Sandstone

90 Ma

Above: Late Cretaceous (Coniacian) paleogeography (87 Ma) during transgression of Mancos Shale

Opposite: Late Cretaceous (Campanian) paleogeography (80 Ma) during transgression of Cliff House Sandstone

80 Ma

70 Ma

Above: Late Cretaceous (Campanian–Maastrichtian) paleogeography (70 Ma) during regression (final) of Pictured Cliffs Sandstone

Opposite: Late Cretaceous–Early Paleogene paleogeography (65 Ma) following regression of Cretaceous Interior Seaway

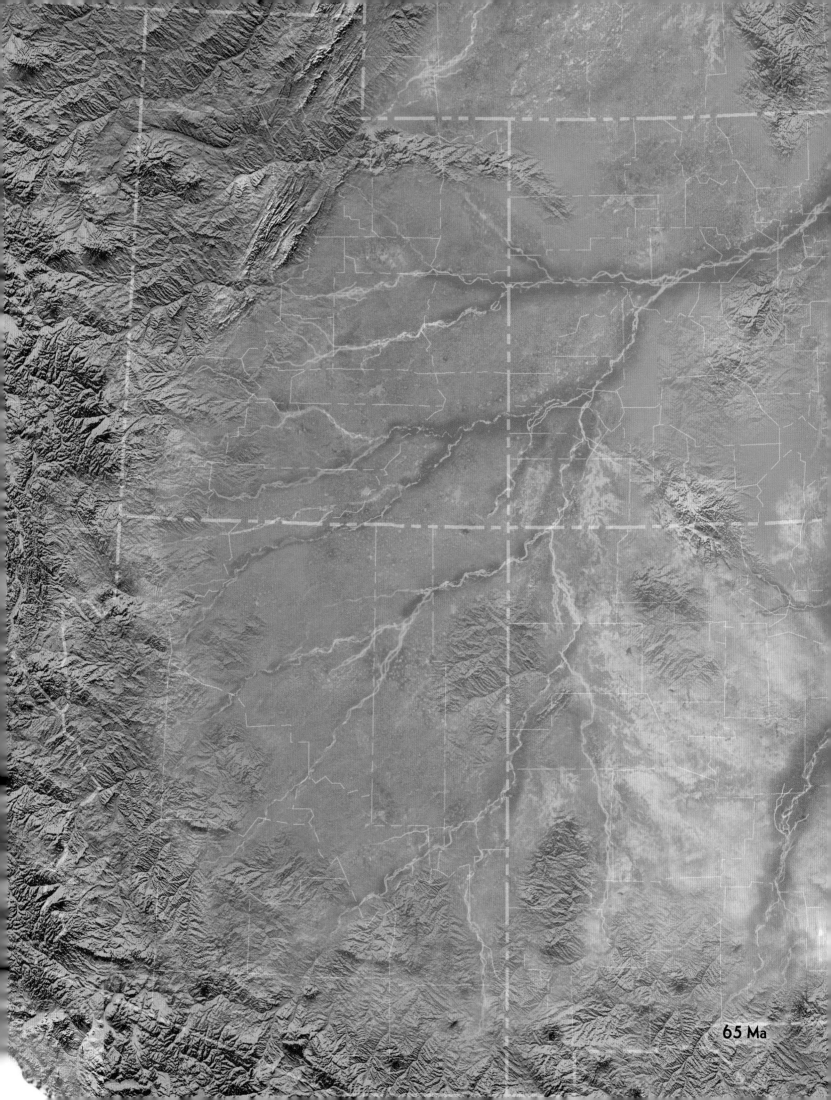

65 Ma

A sandy, shallow marine bottom (with algae) from Rocky Point, Mexico

The adjacent mountains continued to rise during this time, shedding sediments toward the seaway. This sediment precluded deposition of marine carbonate because vast amounts of clastic material overwhelmed conditions needed to produce limestone everywhere except the extreme eastern fringes of the area. Numerous tongues of sand eroded from the mountains into the sea, pinching out within marine shales toward the east. The thickness and extent of the sandstone units varies locally, as do their names, although many units are assigned to the Mesa Verde Group. The sandstone beds form tan to yellow cliffs that contrast with the softer, dark shale. These sandstone deposits extend across most of the plateau region into Colorado, where some future cliff-dwellers would build their homes in overhangs and alcoves created in the sandstone cliffs (Mesa Verde National Park). A total of six or seven cycles of marine transgression (leaving black shale to the west) and regression (preferentially leaving white to gold sandstone to the east) can be defined.

Upper Cretaceous rocks of the Western Interior are famous for the fossils they contain. A marine molluscan fauna with many exquisite and rather large ammonites dominates the marine shale units. Ammonites were cephalopods with an intricate wall structure that is commonly expressed as a swirl-like suture pattern on the shell. Modern relatives to these are called nautiloids and are found today only in the waters off the Philippine Islands. Sharks' teeth are locally abundant, as are bones from marine reptiles. Nonmarine rocks also yield diverse fossil remains including plants, invertebrates, and vertebrates including dinosaurs.

The sandstones of the Mesa Verde Group formed as beach sands were washed in the fluctuating shore of the Cretaceous Interior Seaway. Above is a modern example from the Atlantic coast of Africa.

Toward the end of the Cretaceous, the seaway retreated from the Western Interior for the last time. As the waters withdrew to the east, rivers from the eroding Sevier Highlands spread debris across the entire area and a great alluvial plain extended across what had been the Cretaceous Interior Seaway (the Fruitland Formation and Kirtland Shale in the San Juan Basin, New Mexico). Soon the Rocky Mountains would emerge from these marine shales. While all of these changes were underway, a giant asteroid slammed into a southern section of the Cretaceous Interior Seaway near Yucatan, Mexico. This ten-mile-wide rock from outer space may have caused the demise of the dinosaurs as the impact created global fires and blocked out sunlight. The Cretaceous Period literally ended with a bang!

From the Cambrian through the Late Cretaceous, the Colorado Plateau region had been at or very near sea level. Review the sequence of paleogeographic maps and notice how many times shallow seas covered the vicinity. The withdrawal of the Great Western Interior Seaway signals the last time that ocean water will inundate this area. Within a short time, the Colorado Plateau region will rise from its long existence near sea level and become the high-elevation plateau that reflects its name.

Location of Cretaceous Rocks

Cretaceous rocks generally rim the subbasins of the Colorado Plateau. In Arizona they form the cliffs and mesa tops that comprise Black Mesa and the Hopi mesas. They underlie New Mexico's San Juan Basin, and in southwest Colorado, Cretaceous rocks form the cliffs and redoubts in Mesa Verde National Park. In south-east Utah, they surround most of the Abajo and Henry mountains. North and west of the Henrys, they form the drab, gray badlands on the east side of Capitol Reef National Park including North Caineville Mesa and towering Factory Butte. Cretaceous rocks form the Kaiparowits Plateau in Grand Staircase–Escalante National Monument and the Gray Cliffs near Bryce Canyon National Park as well. They underlie the Wasatch Plateau, rim the San Rafael Swell and then snake eastward in the spectacular Book Cliffs, continuing into De Beque Canyon east of Grand Junction, Colorado. Cretaceous rocks outcrop along the northern Uinta Basin and through southern Dinosaur National Monument and into northwest Colorado.

Chapter Summary

The Cretaceous time period ushered in increasingly wet and humid conditions on the Colorado Plateau. The Sevier Highlands rose in Nevada and supplied abundant sediment to the plateau surface. At the same time, the Cretaceous Interior Seaway flooded the plateau from the east and inundated the landscape with marine mud. Shifts in the position of the shoreline were caused by fluctuations in sediment influx, itself determined by climate, sporadic uplift of the Sevier Highlands, and sea level rise or fall. Swamps were present onshore behind the beaches and great quantities of coal were formed in the rocks. The sea finally withdrew for the last time at the end of the period and left in its wake the "blank canvas" upon which the modern Rocky Mountains and the Colorado Plateau would be formed.

Above: Cretaceous rocks exposed near Lake Powell in Glen Canyon National Recreation Area

Below: The Cretaceous Period ended abruptly when an asteroid about ten miles in diameter collided with Earth near the coast of Yucatan, Mexico.

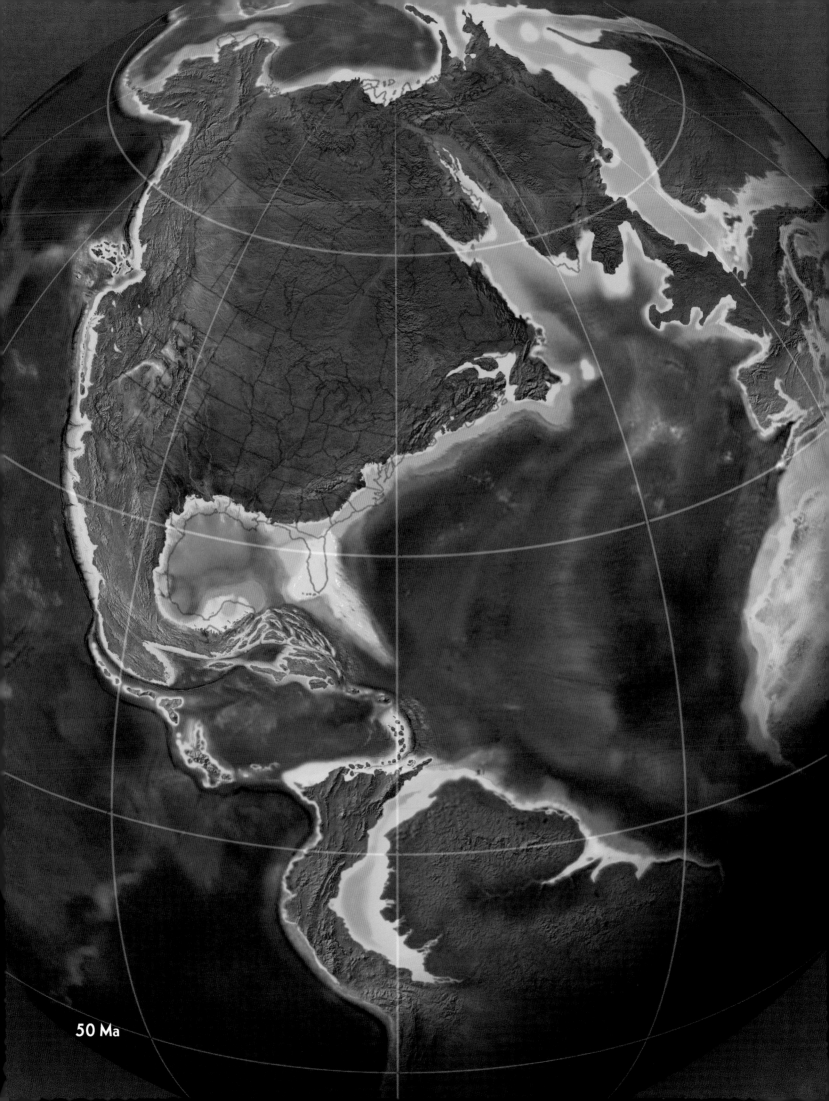

50 Ma

CHAPTER SEVEN
THE MODERN LANDSCAPE EMERGES
CENOZOIC ERA: 65 MILLION YEARS AGO TO THE PRESENT

Long cycles of change define the geologic history of the Colorado Plateau, and throughout time, vast periods of sediment accumulation have melded and merged with equally vast periods of uplift and erosion. These repetitive cycles confirm the idea that the more things change, the more they stay the same. The numerous cycles of Paleozoic transgression and regression and the Mesozoic tug of war between river and eolian environments all foreshadow the dawn of the Cenozoic. And while it might be tempting to view this most recent geologic era as just another predictable cycle of uplift and erosion, it is the very scale of the uplift and the magnitude of the erosion that make this whole story worth telling. In fact, the Colorado Plateau may be Earth's largest mass of uplifted, but relatively undeformed, sedimentary rocks. It is here, during the Cenozoic, that the Colorado Plateau finally attains its formidable and recognizable shape. All of the previous geologic history thus far has merely created the canvas upon which the actual masterpiece is painted.

What significant changes appear! The humid Cretaceous becomes the semiarid Cenozoic. The lowland interior seaway would now become the Rocky Mountains and the Colorado Plateau. A fauna replete with giant dinosaurs and marine reptiles now gives way to species of mammals and birds, some almost as large. Eventually, ice would overcome much of the land in both hemispheres in repetitive cycles of icy glacial advances and thawing meltwater retreats. During this ice age, one species of primate would stand upright on the grasslands of tropical Africa and begin the slow transition from a tree-dwelling leaf eater to a geologist with the ability to decipher much of Earth's history.

20 Ma

Historically, the Cenozoic Era has been divided into two epochs, the longer Tertiary (65 to 2 million years ago), and the much shorter Quaternary (2 million years ago to the present). However, a new subdivision has been recently proposed which splits the Cenozoic into two nearly equal periods, the Paleogene (65 to 23 million years ago) and the Neogene (23 million years ago to the present). The Paleogene Period is further subdivided into three useful and often-used time slices called epochs: the Paleocene (65 to 58 Ma), the Eocene (58 to 38 Ma), and the Oligocene (38 to 23 Ma). The Neogene is also subdivided into the Miocene (23 to 5 Ma),

Opposite: Global view of the Paleogene (50 Ma). The shapes of the modern continents become apparent and mountain building occurs over much of the earth.

Above: Global view of the Neogene earth (20 Ma)

the Pliocene (2 Ma to 10,000 years ago), and the Holocene (10,000 years ago to the present). Seeing the proverbial handwriting on the wall, we use the newer terminology here. The resulting disuse of the terms Tertiary and Quaternary, words that have been used throughout the course of our professional careers, is perhaps a bitter pill to swallow but a milestone worth noting nonetheless. By adopting this new terminology, we become students ourselves in the single most important lesson that our science has taught us—that change is inevitable and the only constant that exists.

Deformation, Uplift, and Erosion: Paleogene Paleogeography

The Paleogene Period ushered in a monumental period of uplift and erosion in the Western Interior of North America. The reasons for this uplift are complex, even controversial among geologists. But perhaps the most favored hypothesis involves the way subduction proceeded along the western margin. Beginning in the Late Triassic and continuing until the Late Cretaceous, the subduction of plates beneath North America was at a relatively steep angle of forty-five to sixty degrees. This steep descent caused uplift and related volcanism to be located very near the plate boundary (as in the Nevadan and Sevier orogenies far west of the Colorado Plateau). Starting in the Late Cretaceous and continuing into the Paleogene, the angle of subduction may have eased to twenty-five degrees or less. This is an unusual angle for a plate to descend but could explain how and why the Rocky Mountains were uplifted almost a thousand miles (1,600 km) inland from the plate boundary. The position of the Rockies so far from a plate boundary makes them one of the most enigmatic mountain ranges on Earth. This far-reaching deformation event is what uplifted the Colorado Plateau as well.

A view of the Waterpocket Fold, Capitol Reef National Park, Utah

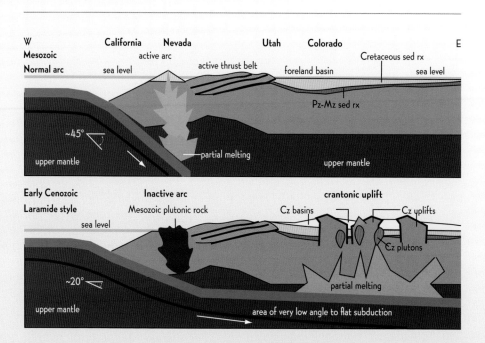

Panel showing how subduction angles beneath North America may have become shallow, causing the uplift of the modern Rocky Mountains and the Colorado Plateau

This specific mountain-building event is called the Laramide Orogeny, which took place from 70 to 40 million years ago. This event not only uplifted the Rocky Mountains but also produced a fair share, if not most, of the uplift of the Colorado Plateau, though geologists vigorously debate the specific timing and frequency of plateau uplift. Historically, uplift of the plateau was thought to have occurred in three discrete pulses during the Cenozoic. But more and more geologists are arguing that the bulk of plateau uplift may have occurred during the Laramide. The pendulum of opinion continues to swing. It is important to note that although all of the Western Interior was uplifted during the Laramide, the Rockies and the Mogollon Highlands were lifted higher than the Colorado Plateau. In the Paleogene, the plateau became a high-elevation basin ringed by even higher mountains. Laramide uplift extended all the way from the Sierra Nevada in the west, under the Great Plains in Kansas and Nebraska to the east, and may have raised the Colorado Plateau as much as 15,000 to 17,000 feet (4,600–5,200 m) at this time—all of this in an area that was covered in marine shales of the Cretaceous seaway just a few million years earlier!

Although rocks in the Rocky Mountains and the Mogollon Highlands were greatly deformed and uplifted, those on the Colorado Plateau experienced only gentle uplift. This type of uplift, called epeirogenic uplift, resulted in only localized folding of the sedimentary rocks, creating structures called monoclines. These features expose spectacular outcrops such as Comb Ridge (Monument Valley), the Waterpocket Fold (Capitol Reef), the Cockscomb (Grand Staircase–Escalante National Monument), and the East Kaibab monocline (Grand Canyon). Sedimentary rocks that once lay flat were compressed and tilted to angles of forty-five degrees or more along narrow (one mile/1.6 km) but relatively long (up to 120 miles/190 km) trends. The folding occurred below the surface, but erosion has exposed the cores of these folds. The monoclines are aligned with faults that had their origins in the Precambrian. The Colorado River in the Grand Canyon cuts deep enough to expose the faulted depths of the East Kaibab monocline.

The Laramide Orogeny also produced a broad sweep of volcanism that moved through time from west to east and back again. This wave of volcanism began to flare up 70 million years ago near the continental edge. Then, as the angle of subduction flattened, the volcanism swept eastward until the end of the Laramide. Volcanism swept back toward the west beginning about 30 million years ago as the subducting plate resumed its steep angle.

This sweep of volcanism may have produced the igneous intrusions that punctured some parts of the Colorado Plateau near the end of the Paleogene. These intrusive events formed the laccolith mountains that dot the plateau landscape today—

1. Flat-lying sedimentary rocks at end of Cretaceous; near sea level
sea level

Cretaceous
Lower Jurassic · Middle and Upper Jurassic
Upper Paleozoic · Triassic · Lower Paleozoic
Precambrian

2. Uplift, monoclinal folding, and dissection during Paleocene-Eocene
sea level

3. Planation, perhaps to fairly smooth surface during Oligocene and Early Miocene
sea level

4. Renewed dissection during Late Miocene and Pilocene to Recent
sea level

Vertical and horizontal scales approx. 3,000 ft

Panel showing the sequential development of Colorado Plateau monoclines and landforms

Originally horizontal strata were deformed during the uplift of the Colorado Plateau. Cottonwood Canyon, Grand Staircase–Escalante National Monument, Utah

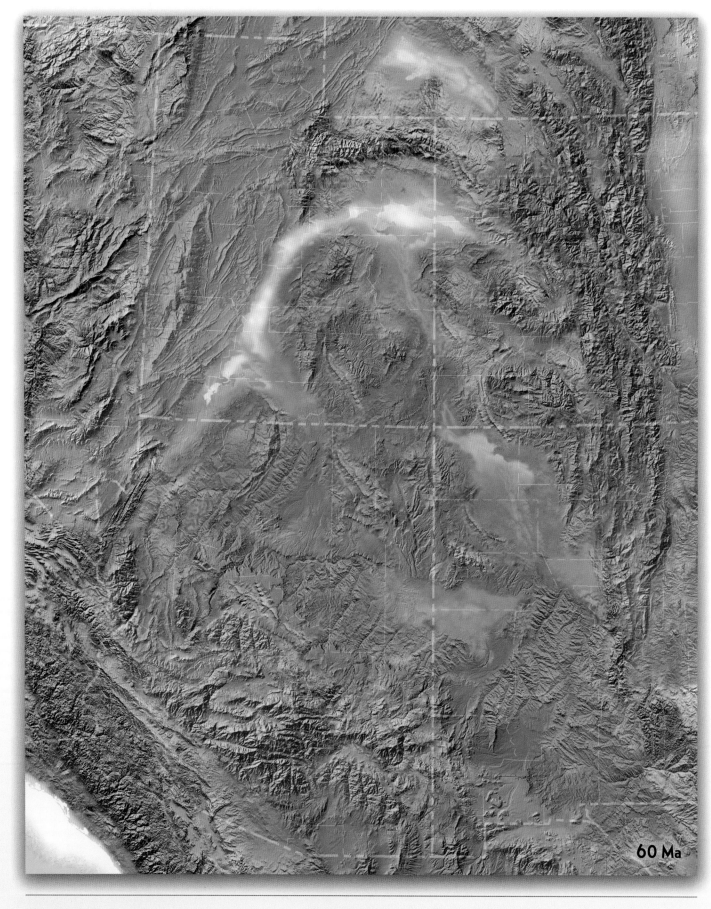

60 Ma

Above: Paleocene paleography (60 Ma). The Paleocene was a time of regional uplift across the Colorado Plateau through folding, faulting, intrusion, and, most importantly, epeirogenic uplift.

Opposite: Early Eocene paleography (50 Ma). Sediments derived from local uplifts were deposited in adjoining basins and formed units such as the Claron Formation in Bryce Canyon National Park and Cedar Breaks National Monument. Lakes formed in some of the deepest basins.

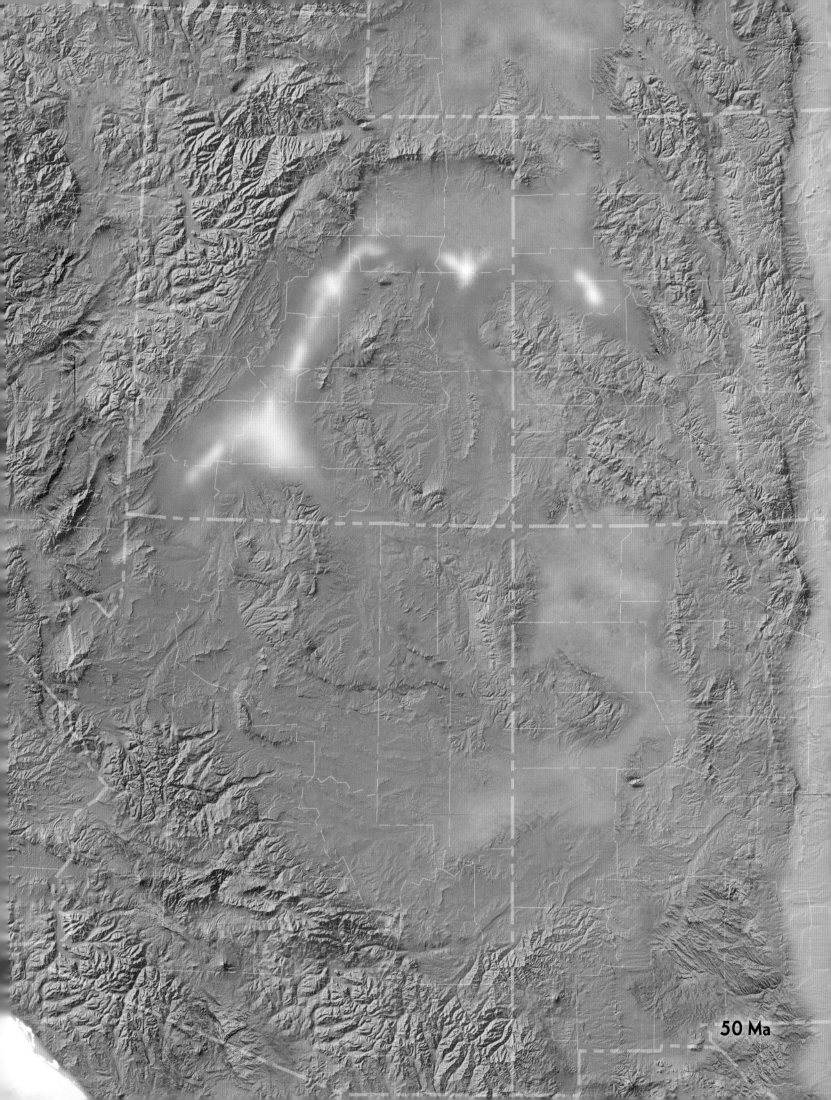

50 Ma

45 Ma

Above: Late Eocene paleogeography (45 Ma). As the Rocky Mountains rose and basins subsided, giant lakes formed on the plateau landscape. The Green River Lake covered parts of three states while scattered sandstone and gravel accumulated in central Arizona and western New Mexico (the so-called "rim gravels").

Opposite: Oligocene paleogeography (30 Ma). Sedimentary rocks of this age are not common and reconstructions are difficult to make. Much of the central plateau may have consisted of smooth topography and was probably the site of hypothetical broad, shallow lakes that reflected the lack of external drainage. Eolian sandstone in the Chuska Mountains suggest a large dune field in that area. Note the incredible expanse of volcanoes across the region—most plateau laccoliths formed during this time.

30 Ma

the Henry, La Sal, Abajo, Sleeping Ute, Carrizo, and Navajo mountains. Laccoliths begin as mushroom-shaped intrusions of magma that horizontally invade layers of sedimentary rock. This causes the rocks to arch upward as the magma thickens. Later erosion has exposed their igneous cores. The heat associated with these intrusive events increases the crust's buoyancy, which may have contributed a bit to the regional uplift.

The La Sal Mountains near Moab, Utah, are an example of a Colorado Plateau laccolith. The sedimentary cover has been eroded off the top of the mountains exposing the crystaline core.

Sedimentation patterns during the Paleogene Period reflect the topographic conditions that existed at the time. Fluvial sand and gravel was derived from both the Rockies (northeast) and the Mogollon Highlands (southwest). Streams flowed from these ranges toward the center of the plateau and formed freshwater lakes in the basins that had warped downward on the plateau's variably folded and differentially uplifted surface. (In fact, the Laramide Orogeny received its name from sediments shed into the Laramie Basin by an adjacent uplift.) These sediments became rock units such as the Claron Formation (Bryce Canyon National Park) and the Flagstaff and Green River formations (northern Colorado Plateau). The Paleogene was a time of hot, humid conditions in the plateau region and some of the rocks contain vast stores of oil shale and tar sands left behind in rocks made carbon-rich by the widespread vegetation and animal life that flourished at this time.

An interesting conjecture related to this Paleogene uplift and sedimentation history, is the idea that parts of the Colorado River system in the Grand Canyon might have been present at this time and flowing northeastward out of the Mogollon Highlands. Ironically, this is opposite the flow direction of this part of the river system today. This scenario, if true, necessitates a much later period of drainage reversal for this part of the Colorado River system, but that is a concern to be dealt with in the Neogene. As remarkable as this initial drainage direction may seem, it is the one part of the Colorado River story that virtually no geologist disputes. The Mogollon Highlands did exist as an Andean-type mountain range, and rivers certainly flowed northeastward out of them. The question remains, however: How much of the modern drainage configuration could be inherited from these Paleogene drainages?

Freshwater lake deposits of the Claron Formation, Bryce Canyon National Park, Utah

Lastly, a widespread surface of erosion formed on the Colorado Plateau and much of the Western Interior near the end of the Paleogene. This surface, called the Telluride surface in Colorado, is preserved beneath various sedimentary gravels and the volcanic rocks of the San Juan Mountains, which began to erupt about 30 million years ago. The uplift and humidity in the Paleogene contributed to this subdued relief. In contemplating the existence of this erosion surface, it is interesting to ponder how much topography might have survived this beveling event. Were Paleogene river configurations positioned by any topographic relief on this surface? Or was the surface of the Late Paleogene landscape completely leveled by erosion and/or buried in gravel? The implications for the alignment of the modern drainage system are huge and cannot be dismissed as the Neogene Period begins.

Extension and Deep Dissection: Neogene Paleogeography

At the start of the Neogene Period, about 23 million years ago, the elevation of the Colorado Plateau was most likely quite high, yet low relative to adjacent landscapes such as the Rocky Mountains and the Mogollon Highlands. Specific details for the plateau landscape at this time are difficult to interpret as very few deposits exist. However, beginning about 17 million years ago, the elevational relationship between the plateau and the Mogollon Highlands to the southwest would be fully inverted. Initially, streams flowed northward from the highlands toward the plateau region, but they were eventually deflected away from the plateau by the emerging escarpment of the Mogollon Rim near Sedona. The evidence of this comes from scattered gravel deposits known as the Beavertail Butte gravels.

The Neogene Period was a time of intense dissection of Colorado Plateau strata. The Goosenecks of the San Juan River are a spectacular example of these canyon-forming processes.

Shortly after deposition of these gravels about 20 million years ago, the Mogollon Highlands began to collapse due to crustal extension. Crust that was originally thickened in the vertical dimension during Laramide compression, became extended in the horizontal dimension and collapsed the mountains. This created the Basin and Range Province south and west of the plateau. Extension of the crust, which may have begun in the Late Paleogene, was strongly underway by about 17 million year ago. Some areas were stretched horizontally more than 100 percent from their previous thickened state. Specific causes for the extension are unclear but possibilities include the gravitational collapse of the Mogollon Highlands, the subduction of the spreading center that once separated the Farallon and Pacific plates, or the escape of built-up heat beneath the overthickened crust. Extension appears to have ceased today in most areas of the Basin and Range in Arizona and Utah but continues in northern Nevada and eastern Oregon.

Paleogene fish from the Green River Formation in Wyoming

15 Ma

Above: Miocene paleogeography (15 Ma). Rocks of this age are rare on the plateau, reflecting a period of erosion. Conglomerate at the foot of Arizona's Mogollon Rim confirms that this feature stood with similar relief and near its present position about 15–20 Ma. This epoch marks the time of Basin and Range collapse southwest of the plateau, and as this happened streams began to nibble away at the plateau edge. However, internal drainage probably still existed on the plateau surface, and numerous ephemeral lakes and dune fields existed in low places.

Opposite: Late Miocene–Early Pliocene paleogeography (5.5 Ma). Scenery and events changed rapidly at this time as the Colorado River became integrated with a route off the plateau and through the Basin and Range to the Gulf of California. Water from the Rocky Mountains began carving the great canyons, and lakes inherited from the previous time of internal drainage now spilled over into headward-eroding canyons.

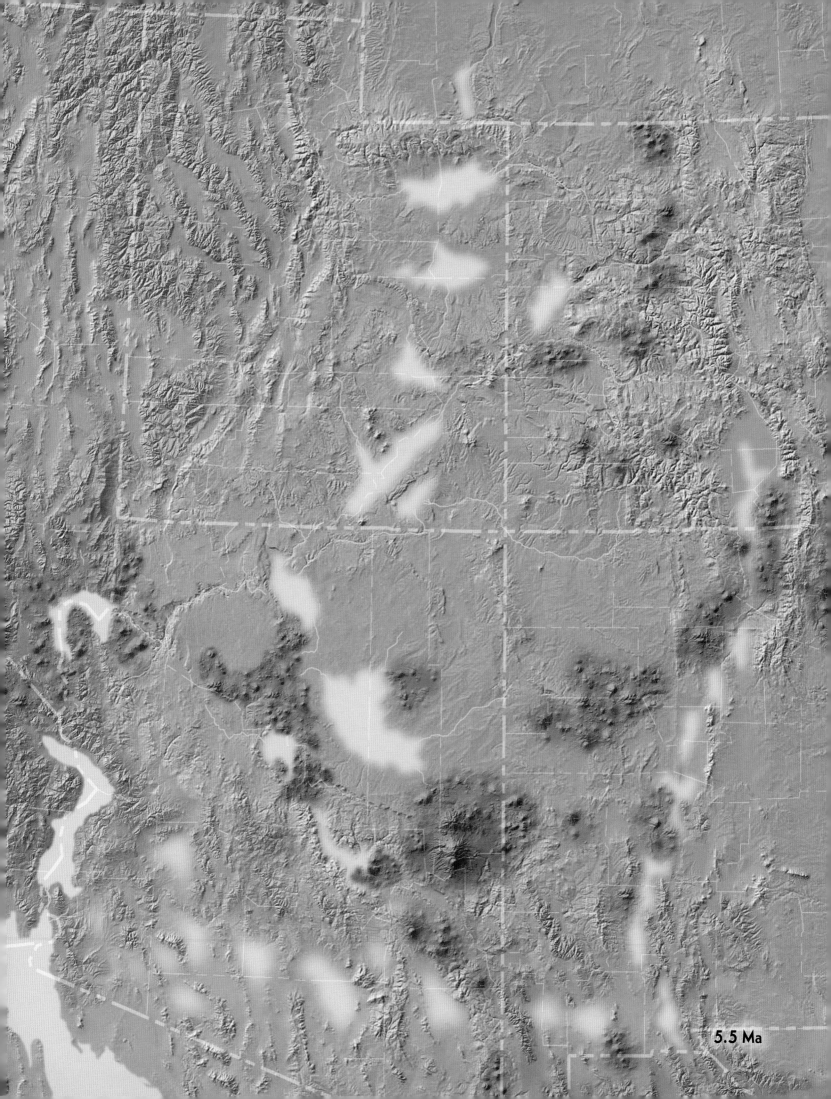

5.5 Ma

The Basin and Range extension appears to be related to a change in the plate boundary that exists in western North America. About 25 million years ago, the margin began to evolve from a long-lived subduction zone with vertical offsets, to a transform boundary with lateral offsets. The San Andreas Fault is the expression of this modern plate boundary. Very little of this extension was able to invade the coherent block of Colorado Plateau crust. However, with the collapse of the Mogollon Highlands, the plateau was finally elevated relative to those areas to the southwest of it, and it attained its present form as an elevated plateau. Arid conditions and the lack of an integrated river system meant that little deposition or erosion occurred on the plateau surface. A spectacular canyon was cut through the eastern Mogollon Rim near the upper Salt River, but there is little evidence of what the Colorado River was doing at this time. Without an integrated drainage system, the plateau probably lacked extensive rivers or deep canyons.

Above: The Colorado River and its tributaries have etched numerous canyons into the uplifted rocks. Glen Canyon National Recreation Area, Arizona

Opposite: Internal drainage into a basin in western Nevada forms this ephemeral lake. Such scenes may have been commonplace on the early Neogene Colorado Plateau.

Neogene streams that coursed across the plateau surface may not have had a single, integrated outlet off of the elevated, saucerlike region. These streams may have drained to internal basins where they could have formed ephemeral lakes, which could evaporate on, or perhaps infiltrate into, the landscape. Any deposits that might have been left from such lakes either never existed or have eroded away. An exception is the Bidahochi Formation in eastern Arizona, but its environment of deposition is still unresolved; it may or may not have been the site of a Neogene lake. While the early Neogene landscape of the plateau was slowly evolving into the present one, a crucial landscape element was missing—deep canyons. Some aspects of the modern drainage configuration may have been in place by this time but the deep dissection that characterizes the modern plateau landscape was not one of them.

By the late Neogene, the lower Colorado River drainage became established as the San Andreas Fault ripped open the Gulf of California. This is a critically important factor in understanding the origin of the Grand Canyon and the Colorado River. Deep dissection of the plateau occurred with the opening of the gulf just prior to 6 million years ago. The opening of the gulf provided an outlet to the sea for the previously disorganized system of rivers on the plateau. Integration of these once disparate river systems was facilitated by the lowering of base level along the newly formed lower Colorado River. The lowering of base level would have caused the Colorado River on the plateau to deepen its track, which in turn would cause it to lengthen its course in the upstream direction. This would facilitate the interception and capture of rivers that previously were not connected to it.

These musings help to illustrate why the timing of plateau uplift has only recently become problematic. The old maxim "no uplift, no canyons" is perhaps better expressed today as "no base-level change, no canyons." Early geologists noted the apparent youthfulness of the deep canyons and invoked a period of recent uplift to explain it. However, a clearer understanding of how and when the Basin and Range was lowered relative to the Colorado Plateau could explain the recent dissection of canyons without invoking recent plateau uplift. Recent plateau uplift (the last 10 million years) has been called into question by some geologists because of the apparent lack of evidence for late Neogene displacements.

In any case, it is likely that a stream located on the plateau surface, and which in part may have originated as a component of the Laramide (northeast) system of drainage, was integrated with the lower Colorado River. This integration event could have been accomplished in any number of ways or combination of ways. A historically popular theory, known as headward erosion and stream piracy, would have occurred just prior to the onset of deep dissection. In this scenario, the ever-deepening lower Colorado River trough caused one of its tributary streams to extend its channel eastward, perhaps along a relatively shallow preexisting channel, until it tapped into the internal drainage system of the upper Colorado River. This would have occurred before the deep dissection of the canyons on a rather subdued landscape since headward erosion is not an efficient process in river systems with steep, localized headwalls.

Alternative scenarios have been proposed. One suggests that the integration event was consummated by the sequential infilling of plateau lakes with sediment that displaced their waters, ultimately resulting in catastrophic overflow of their lowest rims. According to this theory, as water raced downhill to the next basin, it inscribed the course of the Colorado River we see today. One of the spills is postulated to have occurred near eastern Grand Canyon when Lake Bidahochi may have spilled and connected the upper river system with the lower Colorado River, perhaps deepening the Grand Canyon in the process. Yet another idea suggests that integration of the river was accomplished in a series of subterranean caves associated with karst processes. The Grand Canyon is riddled with artesian solution caverns in the Redwall Limestone and, as envisioned, late Neogene groundwater would have been directed under the Kaibab upwarp through these caves. Ensuing cave collapse then caused the surficial integration of the river system.

Massive lava flows spilled into the Grand Canyon and dammed the Colorado River numerous times beginning 630,000 years ago.

Aerial view of Sunset Crater near Flagstaff, Arizona, one of the youngest volcanic features on the plateau

All three theories neatly explain how the many ninety-degree turns seen in the river's modern configuration could form. Each theory suggests that integration was a relatively recent phenomenon. Whatever theory is invoked, the newly integrated Colorado River system eventually cut down rapidly as it came to grade with the lower part of the system at the Gulf of California. Deep, narrow canyons appeared relatively quickly. This downcutting just happened to be across one of the larger uplifts on the Colorado Plateau—the Kaibab upwarp and its eastern limb, the East Kaibab monocline. Through this structure, the Colorado River carved the Grand Canyon. Once the river had an outlet to sea level, tributary streams were quickly integrated into an efficient drainage system that cut the signature landscape element of the region today—canyons. This suggests that the modern Colorado River system as we know it today is less than 6 million years old and that the Grand Canyon is one of the youngest features of that system.

Continued volcanism and faulting has affected the master drainage system in any number of places. Perhaps the most spectacular is in western Grand Canyon where various lava flows filled parts of the canyon and created basalt lava dams and temporary lakes upstream; all beginning about 630,000 years ago. The river ultimately carved through or around these dams, but later flows repeated the process over and over again—as many as thirteen lava dams have been postulated in the Grand Canyon. One of them, the Prospect Dam, would have stood 2,300 feet (700 m) above the Colorado River and a reservoir contained behind it would have stretched upstream nearly to Moab, Utah. At least five catastrophic-outburst flood deposits have recently been documented in western Grand Canyon, suggesting that lava dams did hold water for a time but were destroyed in a geologic instant, perhaps in just dozens or hundreds of years.

Some of the youngest geologic features on the Colorado Plateau landscape are the numerous volcanic fields that can be found on its western, southern, and eastern margins. The San Francisco volcanic field located near Flagstaff, Arizona, began to erupt 6 million years ago but its youngest volcano, Sunset Crater, is less than one thousand years old. The origin of the magma in these volcanoes is unknown, but it may be related to stationary hot spots in the upper mantle or to Basin and Range extension of the crust. If the cause is extension, it actually may indicate the initial stages of rifting the continent's margin into several smaller plates. Stay tuned! The Mount Taylor volcanic field is located near Grants, New Mexico, and recent lava flows and cinder cones are also found on the Markagunt Plateau north of Zion National Park.

Opposite: Present-day geography of the Colorado Plateau. After 1,750 million years of geologic history, over 15,000 feet (4,600 meters) of strata are preserved, enabling us to read and interpret the sequential development of the landscape.

Present day

Another major event late in the history of the region was Pleistocene glaciation. Although it appears unlikely that the Colorado Plateau ever had major amounts of glacial ice, surrounding highlands in the Rocky Mountains underwent significant glaciation during the last 2 million years. What this means for the plateau is that rivers traversing the region probably carried significantly more water during periods when the glaciers were receding. Larger rivers tend to carve deeper canyons. This means that canyon cutting may have operated at a much faster rate during the glacial Pleistocene than it does today.

What seems clear is that the Colorado Plateau region will continue to be one of the more dynamic landscapes on Earth. The major rivers still must cut between 2,000 and 5,000 feet (600–1,500 m) to reach sea level. Most plateau uplands are at elevations between 5,000 and 7,000 feet (1,500 and 2,100 m)—that's a lot of landscape to be removed before the region even begins to approach sea level. As long as the region remains somewhat arid, and as long as the major river systems continue to flow year-round, mesas, buttes, and canyons of the plateau landscape will persist well into the geologic future.

Location of Cenozoic Rocks

Cenozoic rocks are exposed in the centers of basins on the Colorado Plateau. In most places they form ragged, brownish cliffs and multicolored badlands. In Arizona they are restricted to a small area east of St. Johns and into western New Mexico; they also form scattered outcrops along the top of the Mogollon Rim and south of Grand Canyon near Peach Springs. In southern Utah, Paleogene rocks cap the Pink Cliffs from near Cedar City eastward to Bryce Canyon where intricate spires have been recently carved into these lake and stream deposits. Paleogene rocks crop out along the Wasatch Plateau and northward into the Uinta Basin where they form the Roan Cliffs and underlie much of that basin. In northwest Colorado, they underlie the Piceance Basin. In the Uinta and Piceance basins the rocks contain great volumes of oil shale and related petroliferous material (conventional oil). In northwest New Mexico, Paleogene rocks underlie the center of the San Juan Basin, forming brown and tan badlands rich in fossil remains.

Extensive late Paleogene volcanics form the Marysvale volcanic field in southwest Utah and continue west into the Basin and Range. Most of the igneous rocks of the laccolithic mountains (such as the Henrys, Abajos, and La Sals) are late Paleogene in age. The volcanic intrusions in and around Monument Valley and Shiprock are also of this age. The San Juan Mountains of southwest Colorado include Paleogene to Neogene volcanic rocks.

View of Shiprock, New Mexico, a recognizable volcanic landform on the Colorado Plateau

Neogene volcanics are scattered across the margins of the Colorado Plateau. They consist of basaltic lava flows, associated ash and cinder deposits, and large andesite volcanoes. In Arizona they stretch from the North Rim of Grand Canyon, through the San Francisco volcanic field, and eastward to the Mogollon-Datil volcanic field. Sunset Crater, north of Flagstaff, is the most recent volcanic eruption on the Colorado Plateau. Late Cenozoic volcanics are abundant in the Mogollon-Datil and Mount Taylor volcanic fields of western New Mexico; in Utah, volcanics dot the southwest corner of the state and continue northward along the Wasatch Plateau.

Neogene sedimentary rocks are exposed as the Bidahochi Formation in the Little Colorado River valley, the Muddy Creek Formation in the upper Lake Mead region of the Basin and Range Province, and in the Verde Valley of Arizona where limestone, mudstone, and sandstone are present as valley fill deposits of the Verde Formation. Pleistocene sediments are ubiquitous on the Colorado Plateau and are too numerous to list here. Minor glacial deposits are restricted to the Wasatch Plateau and some of the higher volcanic and laccolithic mountains; the nearby San Juan Mountains in southwest Colorado were heavily glaciated.

Erosional landforms like these at Goblin Valley State Park, Utah, dominate the modern landscape of the Colorado Plateau. These are carved into the Entrada Sandstone.

Chapter Summary

The Cenozoic Era is a time of uplift and erosion on the Colorado Plateau. As the Cretaceous seas retreated, the Rocky Mountains and the Colorado Plateau rose through their former floors. Uplifts were separated by basins. Rivers from the Mogollon Highlands to the southwest supplied sediment to plateau lakes. There are few to no deposits in the middle of this era and the history is obscure. But by 17 million years ago, the Mogollon Highlands were down-faulted as the Basin and Range Province formed. Rivers began to run southwest off the plateau, and as the San Andreas Fault deepened the lower Colorado River trough, water was directed there. About 5.3 million years ago, the Colorado River became integrated and through flowing. A deep system of canyons was carved as glacial ice melted in the Rocky Mountains and carried loads of sediment down the rivers.

Hoodoo arch carved into the Entrada Sandstone near Hanksville, Utah

Chapter Eight
Summary of Historical Geology
Grand Canyon and the Grand Staircase

From the bottom of the Grand Canyon to the rim of Bryce Canyon is a straight-line distance of less than one hundred miles (60 km), but within this relatively short distance rocks ranging in age from Lower Proterozoic (1,750 million years ago) to Upper Cenozoic (less than 1 million years ago) are laid out like a giant staircase across the landscape. Clarence Dutton, the eminent protégé of John Wesley Powell, named this feature in the early 1880s when he recognized that rocks from the bottom of the Grand Canyon to the top of Bryce were a gigantic stairstep of sediments with younger ones consistently on top of older ones. Within this giant section of rock, all time periods are represented except those of Ordovician and Silurian age.

Throughout the geologic time covered in this book, the Colorado Plateau region has been a part of the North American continent. The rocks that form the continent's foundation, or basement, actually began to accumulate as sediments and volcanics some 2,000 million years ago, almost half of all earth history. These basement rocks were accreted onto the nucleus of the continent beginning about 1,750 million years ago, during a series of mountain-building events that attached more land to a once smaller North America. Later in the Precambrian, the mountains were worn down flat and sedimentary rocks were deposited across the area. Another period of mountain building resulted in block-shaped mountains across the area, which themselves were later faulted and worn down, in a process that gave birth to the western ocean, a major influence throughout ensuing geologic time.

As these younger Precambrian sedimentary rocks were mostly (but not entirely) removed by erosion, another extensive plain was cut into the landscape. During most of the Paleozoic and Mesozoic, over 475 million years of time, the area of the Colorado Plateau was at or near sea level. A variety of shallow seas and coastal plains covered the region during this time. These areas contained the many depositional systems in which the rock record was deposited. This sedimentary record was dynamic and recorded a number of varied and complex ancient environments. Through subsidence and burial by younger deposits, the sediments became rock and were preserved.

The current excellent, spectacular, and extremely continuous rock exposures across the Colorado Plateau allow us to study the sedimentary rocks in vivid detail. Perhaps nowhere else on Earth are the outcrops so well exposed in such a large area. The rocks began to appear from their long and deep burial about 70 million years ago. This is when the landscape-forming events abruptly and dramatically changed from dominantly deposition near sea level to uplift far above it. It was during this time, called the Laramide Orogeny, that this part of North America became tectonically active and the region was uplifted. Faulting, folding, and

Opposite: Beautifully sculpted landforms in Vermilion Cliffs National Monument, Arizona

Cross-Sections Showing Intervals of Time across the Grand Canyon–Grand Staircase Region

Panel 1: (>1,800 Ma) The deposition and accumulation of a thick stack of sediments and lavas before island arc collision with North America

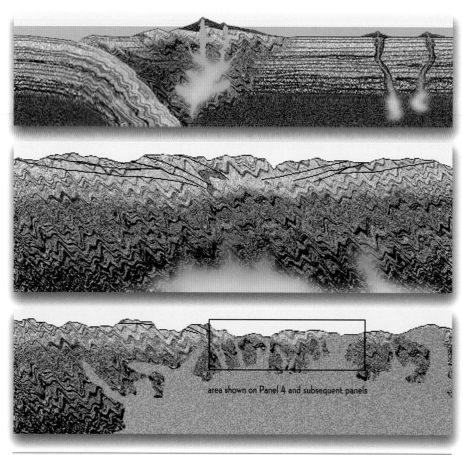

Panel 2: (1,750 Ma) During collision, the rocks in panel 1 were subjected to mountain building, compression, metamorphism, and granite intrusion, which thickened the crust.

Panel 3: (1,750–1,300 Ma) While the rocks were still deeply buried (10 miles [16 km]), granite intrusions continued intermittently.

area shown on Panel 4 and subsequent panels

Panel 4: (1,750–1,300 Ma) This panel is zoomed from panel 3 in the same time interval. All subsequent panels are at this scale of approximately 100 miles (160 km) left to right.

Panel 5: (1,255 Ma) The once deeply buried midcrustal rocks are now on the surface, planed flat to near sea level.

Panel 6: (1,255–740 Ma) Rivers, eolian dunes, and shallow marine settings deposit the Grand Canyon Supergroup rocks (overall thickness more than 12,000 feet [3,700 m]) on the eroded surface. Cardenas Lava erupted approximately 1,100 Ma as dikes and lava flows.

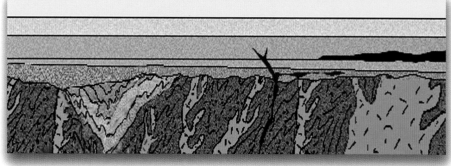

Panel 7: (740–650 Ma) The region experienced another episode of uplift, faulting, and tilting. The Grand Canyon Supergroup was left deformed upon the landscape.

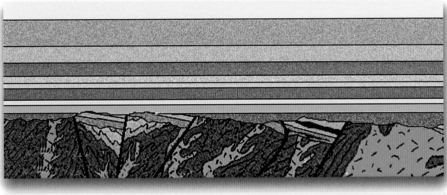

Panel 8: (650–525 Ma) The Supergroup rocks were highly eroded in most areas and left as isolated fault block remnants everywhere else.

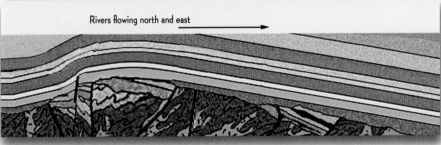

Panel 9: (525–70 Ma) Panel showing most of Phanerozoic history when the plateau region was near sea level and was subsiding to accumulate between 14,000 and 18,000 feet (4,300–5,500 m) of sediment.

Rivers flowing north and east

Panel 10: (70–40 Ma) The uplift and creation of the Rocky Mountains and the uplift and formation of monoclines on the Colorado Plateau. Youngest layers on the plateau began to erode.

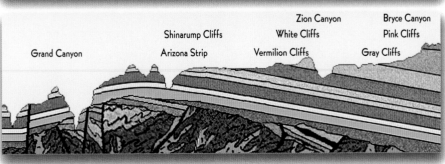

Panel 11: (10–5 Ma) Panel showing the most recent episode of erosion when the Grand Canyon began to take its modern shape. Some geologists think that parts of the canyon may have begun to form during the time in panel 10, but this panel shows when the modern Colorado River became integrated.

Grand Canyon   Shinarump Cliffs   Zion Canyon   Bryce Canyon
Arizona Strip   White Cliffs   Pink Cliffs
Vermilion Cliffs   Gray Cliffs

Panel 12: (2–0 Ma) The modern landscape from the Grand Canyon through the Grand Staircase. The total thickness of rock is between 14,000 and 18,000 feet (4,300–5,500 m), but the change in elevation is only 7,000 feet (2,100 m). The discrepancy is resolved when taking into account the regional dip on strata of 1 to 2 degrees to the north.

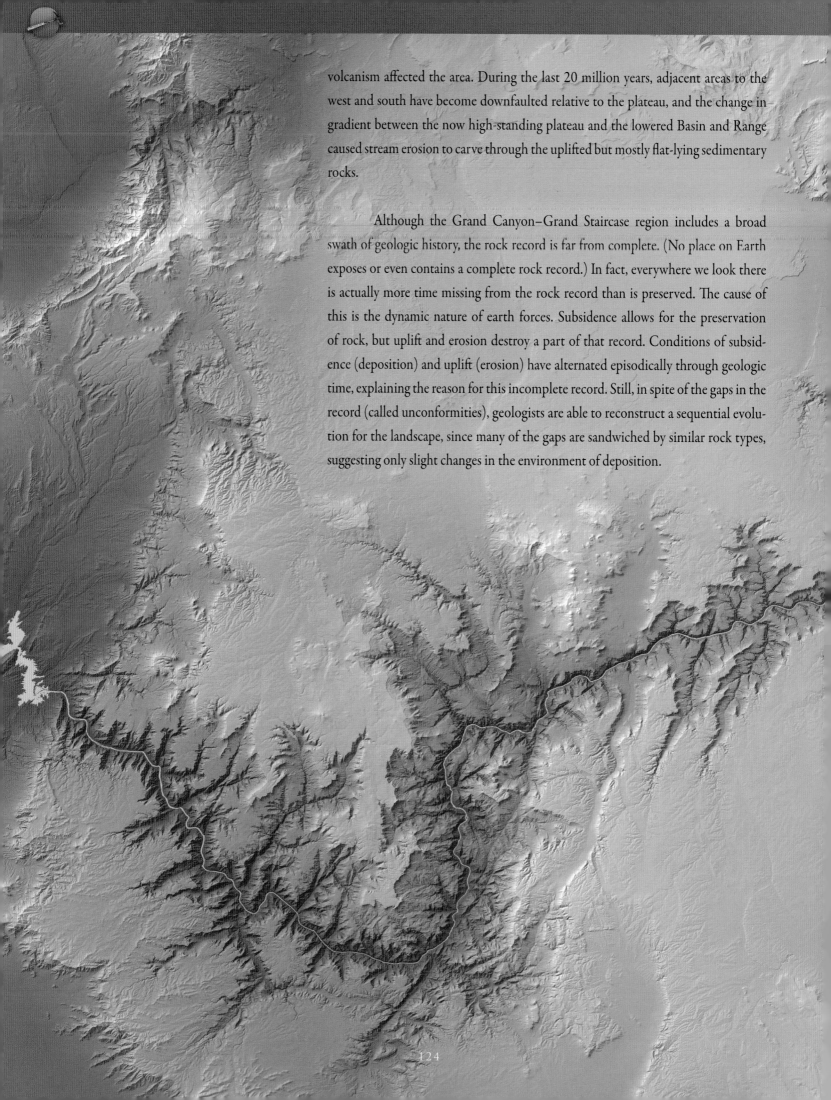

volcanism affected the area. During the last 20 million years, adjacent areas to the west and south have become downfaulted relative to the plateau, and the change in gradient between the now high-standing plateau and the lowered Basin and Range caused stream erosion to carve through the uplifted but mostly flat-lying sedimentary rocks.

Although the Grand Canyon–Grand Staircase region includes a broad swath of geologic history, the rock record is far from complete. (No place on Earth exposes or even contains a complete rock record.) In fact, everywhere we look there is actually more time missing from the rock record than is preserved. The cause of this is the dynamic nature of earth forces. Subsidence allows for the preservation of rock, but uplift and erosion destroy a part of that record. Conditions of subsidence (deposition) and uplift (erosion) have alternated episodically through geologic time, explaining the reason for this incomplete record. Still, in spite of the gaps in the record (called unconformities), geologists are able to reconstruct a sequential evolution for the landscape, since many of the gaps are sandwiched by similar rock types, suggesting only slight changes in the environment of deposition.

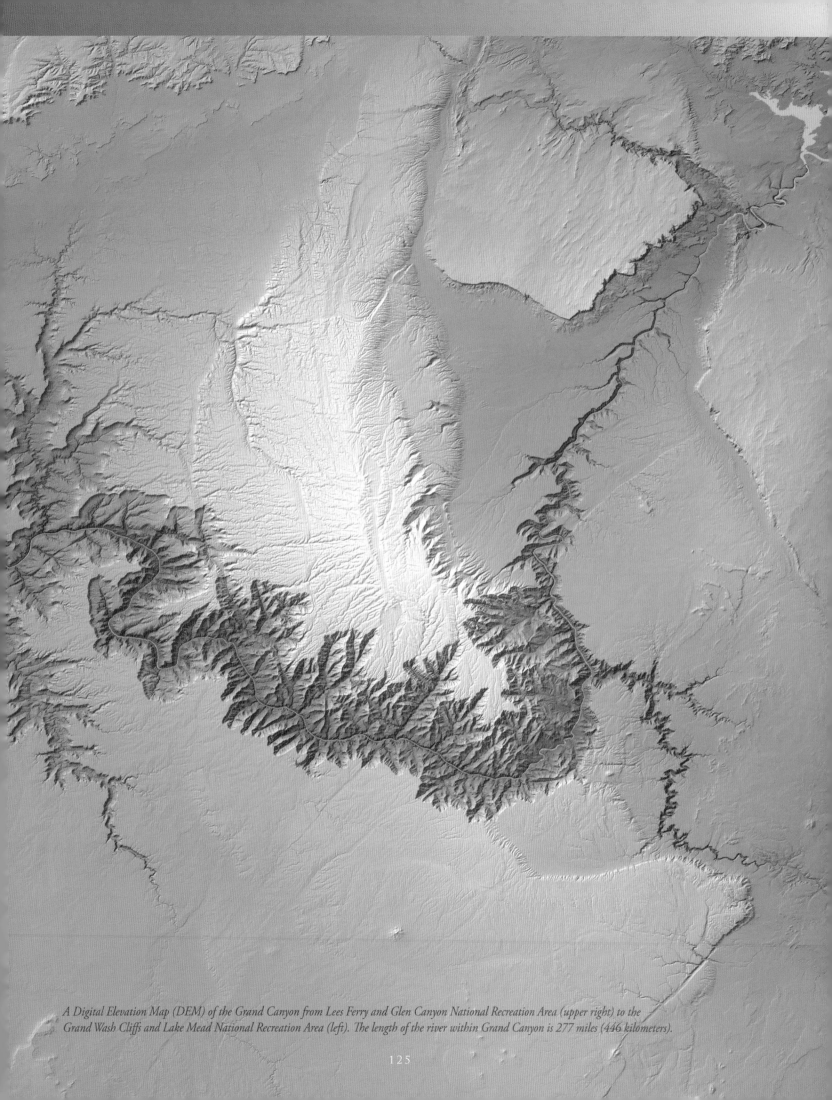

A Digital Elevation Map (DEM) of the Grand Canyon from Lees Ferry and Glen Canyon National Recreation Area (upper right) to the Grand Wash Cliffs and Lake Mead National Recreation Area (left). The length of the river within Grand Canyon is 277 miles (446 kilometers).

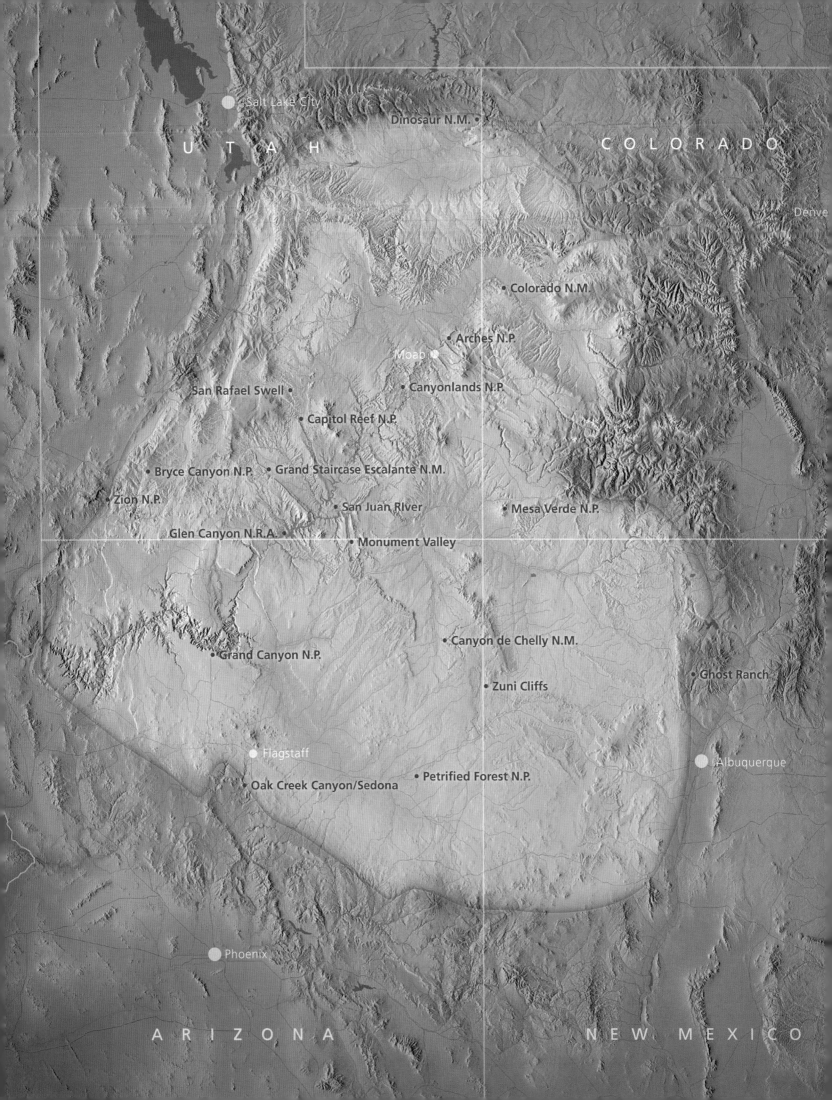

CHAPTER NINE
WHERE TO SEE THE ROCKS

The Colorado Plateau is an excellent place to see rocks and to learn about earth history. Listed here are the best places we know of for viewing the colorful scenery and experiencing fantastic geology. The order of the list wraps around in counterclockwise fashion, starting in Sedona, Arizona, and finishing at—where else?—the Grand Canyon. The list combines the greatest scenery, ease of access, and a wide exposure of rock history on the Colorado Plateau. A visit to these places will provide a great demonstration of the geologic history presented in this book.

The Mogollon Rim and Oak Creek Canyon, Sedona, Arizona

Oak Creek Canyon and the Sedona area occupy a spectacular scenic notch carved into the southern edge of the Colorado Plateau. The area is dissected into beautiful red rock scenery that consists of numerous cliffs, mesas, buttes, spires, and slickrock. The growing

city of Sedona provides first-class amenities that make geologic exploration more palatable to the pampered, but a wealth of national forest wilderness areas and trails beckons the hardy as well. The stately cliff line above Sedona is called the Mogollon Rim (pronounced MUG-ee-on), which rises almost 3,000 feet (900 m) above the Verde Valley. This dramatic change in elevation triggers copious summer rain and winter snow, which support a lush coniferous forest on the rim and verdant riparian trees within the canyon—all in wondrous contrast with the red and orange sandstone formations above.

A complete Permian section more than 2,000 feet (600 m) thick is exposed in Oak Creek Canyon and on adjacent Wilson Mountain. The inner walls of lower Oak Creek Canyon expose the cross-bedded Esplanade Sandstone, one of the oldest of the plateau's eolian deposits and part of the Supai Group. The overlying tree-covered slopes are made up of the fluvial, dark-red Hermit Formation, which forms the broad terrace on which the city is built. The reddish-orange cliffs, buttes, and spires for which Sedona is so famous are composed of the Schnebly Hill Formation, containing some of the finest examples of interbedded eolian and coastal plain deposits found anywhere on Earth. The towering 1,000-foot-high (300-m) cliffs above consist of the eolian Coconino Sandstone and marine Kaibab Formation. Basalt flows that emanated from numerous fissures and volcanoes 15 to 6 million years ago, cap the Mogollon Rim.

Above: Exposures of the Permian section near Sedona, Arizona. Bell Rock (center right) comprised of the Schnebly Hill Formation.

Opposite: Map of the Colorado Plateau showing twenty prime locations to observe the geologic history of the region.

All four gateways to the area—Oak Creek Canyon from the north, Arizona Highway 89A from the west, Arizona Highway 179 via Interstate 17 from the south, and Schnebly Hill Road (high-clearance vehicles only) from the east—provide spectacular approaches to the region. All are flanked by many well-marked, well-maintained hiking trails that lead into the surrounding wilderness areas. The Sedona area provides some of the finest hiking trails to be found anywhere, most maintained by the U.S. Forest Service, some by the Arizona State Parks system, and some by the City of Sedona. Bicycles and horses are welcome on many of the trails.

Petrified Forest National Park, Arizona

The multihued mudstone and sandstone deposits of the Chinle Formation dominate this national park, created in 1906 and enlarged in 2006. The Chinle is exposed in intricate and eerie badlands that weather to reveal the world's greatest concentration of petrified logs. These great stone logs were once stately conifers that bordered the banks of Triassic streams. Geologic evidence suggests that some of these streams had their headwaters in the southern Appalachian Mountains. Volcanic eruptions from the south and west created havoc in this subtropical river system and may have actually killed the trees when they were inundated with layers of wind-borne ash. The dead trees fell into the streams during floods and were brought to their present resting place within the modern park as giant logjams. Continued volcanic eruptions quickly buried the logjams, preserving them from decay and also incarcerating them within layers of silica-rich ash. The silica molecules released from the decomposed ash grew within voids in the wood and ultimately turned the logs to stone.

Petrified logs (above) with exquisite detail of preservation are found in the Chinle Formation in Petrified Forest National Park (below).

Petrified Forest National Park also contains a part of the Painted Desert that continues northwest toward Tuba City and the Echo Cliffs. The colorful Moenkopi and Chinle formations give the desert its name; it is defined by their outcrop exposure. This broad stretch of colorful badlands is one of the Colorado Plateau's iconic landscapes. The colors result from the iron oxides in the sand and clay layers.

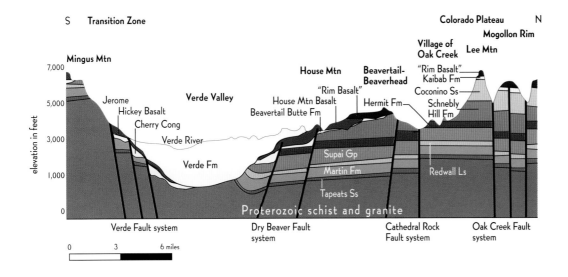

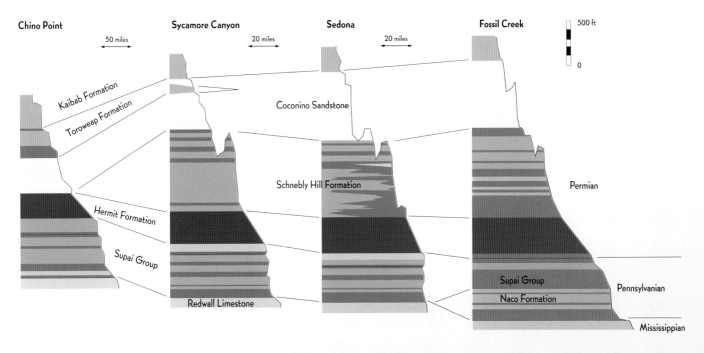

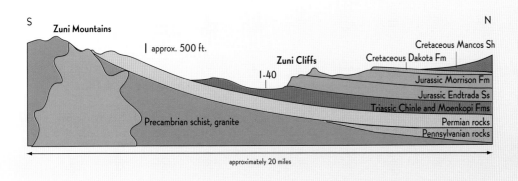

Top: *Cross section of the Verde Valley from Mingus Mountain (left) to the Mogollon Rim (right) showing the rock units, structure, and major landscape elements of the area*

Center: *Correlation of the Permian rock units from Chino Point near Seligman (left) to Fossil Creek southeast of Sedona (right)*

Bottom: *Cross section south to north from the Zuni Mountains through the Zuni Cliffs, showing rock exposures from Precambrian to Cretaceous age*

Zuni Cliffs and Ghost Ranch, New Mexico

The Zuni cliffs are a surprising outcrop along Interstate 40 from the Arizona–New Mexico border, east of Grants. This is approximately a ninety-mile (144-km) drive one-way. Beginning in the west at the state line, deep-orange Jurassic sandstone is exposed as cliffs. These strata dip below the roadway toward the east in a giant syncline that brings the overlying Morrison Formation to road level. Gallup is built upon overlying Cretaceous rocks, and the McKinley Coal Mine is located a few dozen miles north. In a spectacular road cut east of Gallup, the Nutria monocline brings the bright-orange Jurassic sandstone back to the surface, where it is exposed in the Wingate cliffs. These served as the type section for the Wingate Sandstone—until it was determined that this strata actually belonged to the younger, but almost identical, Entrada Sandstone. These cliffs are seen for many miles but eventually give way to volcanic rocks around Mount Taylor near Grants. The McCartys lava flow east of Grants is only about 3,800 (+ 1,200) years old. Beyond Budville, the excellent exposures of contorted Entrada Sandstone give way to the underlying Chinle Formation.

Ghost Ranch lies in north-central New Mexico, northwest of Santa Fe. This area, with its colorful array of Triassic, Jurassic, and Cretaceous rocks, is located at the eastern boundary of the Colorado Plateau where it borders the Rio Grande rift. It was here in 1947, that a rich quarry of Coelophysis dinosaurs was unearthed from within the Chinle Formation. These rather early small dinosaurs were apparently killed in a flood and washed into an eddy or pond where their skeletons were preserved. Nearby, the Echo Amphitheater on U.S. Forest Service land contains an exposure of Entrada Sandstone in a beautiful alcove. A short trail leads into the area.

Ghost Ranch, New Mexico, where strata from Triassic to Jurassic age are exposed

Canyon de Chelly National Monument, Arizona

Canyon de Chelly and its sister canyon, Canyon del Muerto, form a gem of a national monument in the northeastern corner of Arizona on the Navajo Indian Reservation. The sheer canyon walls expose giant cross-beds of the De Chelly Sandstone, which formed in the dunes of a Permian desert and was initially named for the excellent exposures in this canyon. The De Chelly was one of the earliest of the many eolian deposits that grace the plateau landscape. Atop the canyon's rim sits the Shinarump Member of the Chinle Formation of Triassic age. The contact between these two formations represents an unconformity or gap in the rock record that spans almost 50 million years. The Shinarump contains much gravel-size debris, documenting its origin in vigorous braided streams that came from the mountains to the south and east. Don't miss the walk down the spectacular White House Trail, which provides access to the rocks, the canyon floor, and an outstanding cliff dwelling. About

Canyon de Chelly National Monument is the type section for the De Chelly Sandstone, which was deposited in the dunes of a Permian desert.

one-quarter of a mile (400 m) down the trail, look for a Triassic-age slot canyon that was cut into the De Chelly Sandstone some 225 million years ago. It is filled with the coarse river deposits of the Shinarump Member. This is a once-in-a-lifetime exposure not to be missed by the geologist. A drive to the viewpoint at Spider Rock will also reveal how once loose sand can be cemented and eroded into a magnificent spire.

Monument Valley, Arizona–Utah Border, and the Canyons of the San Juan River

Monument Valley is a Navajo Nation Tribal Park and is inhabited by Navajo herders and craftspeople. Check with the tribal park before venturing into the backcountry here. The area is best viewed from the numerous paved and graded roads that traverse the area. Monument Valley may be the finest example of butte and mesa topography anywhere on Earth. Along U.S. Highway 163, the skirts of the buttes consist of the deep-red, fluvial deposits of the Permian Organ Rock Formation. The shafts of the buttes are composed of the eolian De Chelly Sandstone. The thin caps consist of Triassic Moenkopi Formation with distinctive topknots of the cliff-forming Shinarump Member of the Chinle Formation. Monument Valley stands on the flanks of the Monument upwarp and the rocks dip to the west, south, and east off of its crest, such that younger Triassic, Jurassic, and Cretaceous rocks are exposed in those directions.

Three-part diagram showing the sequential development of a typical Colorado Plateau landscape. The carving and growth of canyons (left) causes mesas (center), buttes, and spires (right) to form, such that mesas erode into buttes, which in turn erode into spires. The canyons on the left are typical of Canyon de Chelly, and the buttes on the right can be seen in Monument Valley.

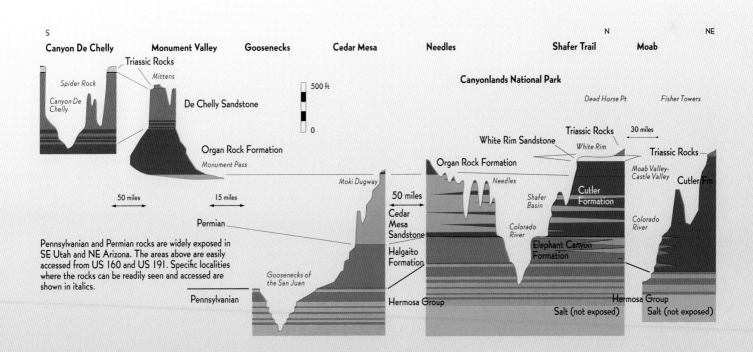

Cross sections showing the relationships between the Pennsylvanian and Permian rock units in five areas on the Colorado Plateau. Note how the rock units in Canyonlands and Cedar Mesa (right) underlie those in Canyon de Chelly and Monument Valley.

To the north, the older Permian and Pennsylvanian rocks are exposed in Goosenecks State Park and adjacent Cedar Mesa. The area is full of interesting side trips. One of the biggest surprises is driving up the switchbacks of the Moki Dugway to Cedar Mesa and Muley Point. This amazing viewpoint sits 6,400 feet (1,950 m) above sea level and takes in a spectacular view of the San Juan River canyons and Monument Valley beyond.

The Goosenecks of the San Juan River are carved into Pennsylvanian rocks belonging to the Hermosa Group. This is the finest example of an entrenched meander to be found anywhere—the San Juan has maintained its meandering course as it carved through resistant marine limestone. If you have the time and have made proper arrangements, a float trip through the canyons of the San Juan offers an outstanding geologic experience. The trip begins at Bluff, Utah, in Jurassic rocks exposed in an open, colorful valley. The river then cuts through the Comb Ridge monocline, consisting of steeply dipping Jurassic, Triassic, and Permian rocks, and marking the east limb of the Monument upwarp. The canyon walls then rise as the river plunges through the center of the Raplee anticline. The canyon opens up once again through the Mexican Hat syncline, but most trips continue downstream through the Goosenecks where the limestone rocks appear again. The river trip ends eighty-six miles (138 km) downriver at the point where the Cedar Mesa Sandstone dives beneath the river as the river flows into Lake Powell.

Another dramatic excursion is the loop through Valley of the Gods (a small-scale Monument Valley) and the short detour to Mexican Hat Rock, an impossibly balanced rock formation. Each of these areas exposes lower slope-forming red rocks of the Permian Halgaito Formation and the overlying yellow-orange cliffs of the Cedar Mesa Sandstone.

A view of the spectacular Moki Dugway, placed on a lower slope of Halgaito Formation and blasted into an upper cliff face of Cedar Mesa Sandstone. The seemingly impossible-to-construct road was built to haul uranium from White Mesa to the mill at Mexican Hat during the Cold War.

Below: The Goosenecks of the San Juan with a meander ratio of 3 to 1 (three miles by river yields only one mile of straight-line distance traveled to the west). The upper slope is composed of the Honaker Trail Formation with a lower gorge cut into the Paradox Formation.

Mesa Verde National Park, Colorado

Located between Durango and Cortez, Colorado, Mesa Verde National Park was created in 1906 chiefly to protect the most spectacular prehistoric cliff dwellings in the Southwest. In the process, some pretty amazing scenery was preserved as well. The mesas and canyons that harbor the cliff dwellings are composed of Cretaceous rocks and tell a fascinating story of how the Cretaceous Interior Seaway fluctuated back and forth some 90 to 75 million years ago. Many rock formations are seen in the park starting with the dramatic climb up to the mesa top. These rocks belong to the Mancos Shale and are prone to slippage and mass wasting making them a bane to the park's road crew. This thick shale represents the deepest phase of the ancient seaway when its shoreline had advanced perhaps 150 miles (240 km) to the west. Once on top of the mesa, the road travels along the alternating tan cliffs and gray shales of the Mesa Verde Group, deposited along the fluctuating shoreline of the seaway.

Alcoves used in the construction of the cliff dwellings were formed within the beach sands of the Cretaceous seaway.

The Mesa Verde Group is about 1,200 feet (370 m) thick in the park. A long tunnel has been cut into the lowermost section (called the Point Lookout Sandstone), which represents the beach sands of the regressing sea. The concentration of ruins is farther down the road and within the younger Cliff House Sandstone. The two sandstone units are virtually indistinguishable but are separated by the Menefee Shale. The Menefee represents a moment in time (perhaps 1 to 2 million years) when the sea regressed to the northeast to expose a fluvial plain, replete with trees that left petrified wood and leaf impressions. The Cliff House Sandstone marks the next transgression of the sea over these fluvial deposits. Fossils on top of the Cliff House Sandstone indicate an age of about 78 million years. The variable layers of sand and shale, which respectively enhance or retard groundwater flow, are what cause groundwater to seep out of the rocks at Mesa Verde, ultimately allowing the alcoves to form. The combination of shelter and water made these alcoves attractive locations for settlement. One must wonder if the ancients who lived here 1,000 years ago could possibly know that their high-perch paradise was within the sands of a Cretaceous beach.

Colorado National Monument, Colorado

Located near Grand Junction, Colorado, this is one of the best places to see evidence of the Ancestral Rocky Mountains. This range popped up about 315 million years ago and was finally buried at this location about 225 million years ago. Along the park's scenic drive, you'll see the colorful cliffs of Wingate Sandstone and Kayenta Formation, both of Jurassic age. Underlying these red cliffs is a thin veneer of the Triassic Chinle Formation. This sits on the dark Precambrian basement rocks creating an unconformity that represents about 1,500 million years of time. During the

Outcrops of Jurassic Wingate Sandstone and overlying Kayenta Formation overlooking the Grand Valley, Colorado National Monument. These sandstones covered the flanks of the Ancestral Rocky Mountains.

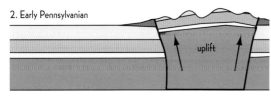

1. Late Mississippian

Mississippian limestone
Devonian sedimentary rocks
Cambrian sandstone
Precambrian continental crust

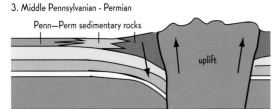

2. Early Pennsylvanian

uplift

3. Middle Pennsylvanian - Permian

Penn—Perm sedimentary rocks

uplift

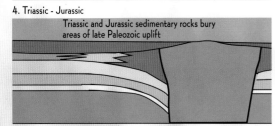

4. Triassic - Jurassic

Triassic and Jurassic sedimentary rocks bury
areas of late Paleozoic uplift

*Above: Panel diagrams showing (top to bottom)
the sequential growth, development, erosion,
and burial of the Ancestral Rocky Mountains
as displayed at Colorado National Monument.*

*Below: Split Mountain, Dinosaur National
Monument*

Pennsylvanian and Permian time periods, this area was the center of a vast mountain range that was uplifted and stripped of its former cover of Paleozoic sedimentary rock. Essentially, all of the strata exposed today in the Grand Canyon were removed from here during the end of the Paleozoic. Similar geology is exposed about twenty miles (32 km) south in Unaweap Canyon, perhaps an ancient, abandoned course of the Gunnison or Colorado River. And farther southeast is Black Canyon of the Gunnison National Park, where the basement rocks are exposed on a grand scale. Here they were not covered until deposition of the Entrada Sandstone, some 60 million years after those at Colorado National Monument.

Dinosaur National Monument, Utah-Colorado

Located in northeast Utah and northwest Colorado, Dinosaur National Monument is a little-visited jewel. A giant dinosaur bone bed, uplifted to near vertical, is displayed in marvelous detail. The bones are the remains of Jurassic dinosaurs that may have been stranded around an evaporating water hole, only to die a horrid death and become preserved within the beds of the Morrison Formation. Most of the monument contains no roads, but two large rivers provide excellent raft access. Near the southwest entrance to the park, well-named Split Mountain is cut in half by the Green River and exposes rocks from Pennsylvanian to Cretaceous age. Split Mountain is an anticline that was once buried under the bed of the Green River then breached by continued cutting by the river. The eastern portion of the park contains the Yampa River, exposing the Pennsylvanian-age Morgan Formation and the Permian Weber Sandstone, both of which document the deposition of sediments that were shed from the Ancestral Rockies.

San Rafael Swell, Utah

The San Rafael Swell in southeastern Utah is one of the most colorful and scenic tracts of land not belonging to a unit of the National Park System. The swell is huge and shaped like a giant kidney bean when viewed from space. This shape reflects its uplift history, which raised the strata here into a broad dome during the Laramide Orogeny, some 40 to 70 million years ago. The dome is dissected on its top and exposes a core of surreal rock outcrops beginning with the Permian White Rim Sandstone (eolian) and Kaibab Formation (shallow marine). These white deposits are capped by the entire section of mostly red Triassic and Jurassic sedimentary rocks, making this an excellent place to view a large section of the plateau's stratigraphy. The San Rafael Swell contains many intricate canyons, imposing buttes, and long lines of cliffs that may take days to even partially explore by bike, foot, or four-wheel drive. The eastern margin of the San Rafael Swell exposes a stunning example of a steep, Colorado Plateau monocline—some of the strata dip at almost ninety degrees. This is best viewed while traveling west on Interstate 70 from Green River in the morning or east from Salina in the afternoon. The highway bisects the uplift as it snakes its way toward numerous scenic pull-outs that provide wondrous views.

Tilted Jurassic strata on the eastern flank of the San Rafael Swell, Utah

Capitol Reef National Park and the Burr Trail, Utah

Surprising Capitol Reef is a wonderland of rocks in Wayne County, Utah. It was named for both a dome of Navajo Sandstone that resembles the dome of the Capitol in Washington, D.C., and a line of cliffs that form the eastern edge of the Water-pocket Fold. These cliffs were a barrier to travel in the early days and any such barrier was often called a reef. This fold marks the trend of yet another spectacular Colorado Plateau monocline. A wonderful view can be had at the Goosenecks Overlook of Sulfur Creek, just off Utah Highway 24. Beneath your feet are the sandstone ledges of the Moenkopi Formation, formed in the deltas at the mouth of an extensive Triassic river system. The canyon below cuts through the Moenkopi and into the marine Permian Kaibab Formation. In the distance to the east, lie the cliffs of the Waterpocket Fold, composed of the basal Chinle Formation and capped by sheer orange cliffs of Wingate Sandstone, the brown ledges of Kayenta Formation, and the golden domes of Navajo Sandstone. A trip down the western side of this uplift on the ten-mile (16-km) Scenic Drive is a trip never forgotten. The unpaved Notum Road, fit for passenger vehicles, leads to the south along the eastern side of the fold. This road takes the curious traveler far off the beaten path to the southern part of the park and the Burr Trail, another dirt road that climbs the Waterpocket Fold in a series of spectacular and (for some) hair-raising switchbacks. At the top is another magnificent view to the east across the monocline, toward the drab-colored but spectacular Cretaceous

Waterpocket Fold along the Burr Trail, Capitol Reef National Park, Utah. Jurassic rocks are exposed in the foreground and middleground, and Cretaceous rocks in the background.

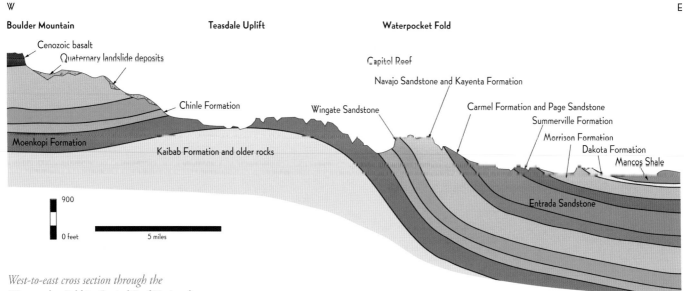

W   E

Boulder Mountain   Teasdale Uplift   Waterpocket Fold

Cenozoic basalt

Quaternary landslide deposits

Capitol Reef

Navajo Sandstone and Kayenta Formation

Chinle Formation

Wingate Sandstone

Carmel Formation and Page Sandstone

Summerville Formation

Morrison Formation

Dakota Formation

Mancos Shale

Moenkopi Formation

Kaibab Formation and older rocks

Entrada Sandstone

900

0 feet   5 miles

West-to-east cross section through the Waterpocket Fold in Capitol Reef National Park, Utah. The fold exposes some of the most spectacular Colorado Plateau scenery.

Delicate Arch in Utah's Arches National Park has formed within the Entrada Sandstone, an eolian dune deposit of the Jurassic Period.

cliffs and igneous laccoliths of the Henry Mountains. The road continues west of the park where it bisects the Circle Cliffs within Grand Staircase–Escalante National Monument, through Long Canyon (carved into the Wingate Sandstone), and eventually continues back over Boulder Mountain to Capitol Reef National Park.

Arches National Park, Utah

The world's greatest concentration of natural stone arches has been carved into the Jurassic Entrada Sandstone just north of Moab, Utah. The arches range from youthful, jagged holes in blocks of sandstone such as at the Parade of Elephants, to older, elaborate spans such as Landscape Arch and Delicate Arch. The rock strata at Arches have been tilted and fractured, but not by the normal means of a Colorado Plateau monocline. The area is part of the salt anticline region that is underlain by rich salt deposits in the Pennsylvanian Hermosa Group. As the Colorado River established its course in the region, water penetrated the Hermosa and dissolved the salt away. The overlying rock layers have been undercut, faulted, and fractured because of the removal of salt at depth. Two parallel ridges dip away from the Salt Valley depression in the center of the park. Each contains hundreds of fractured rock fins of Entrada Sandstone, where the arches are located.

Fins create the background for the arches but the stratigraphy explains why it happened here. Most of the arches are found above the contact of two different types of deposits, an upper pale-orange, cross-bedded sandstone (the Slickrock Member of the Entrada Sandstone) and a darker-red sandstone and mudstone below (the Dewey Bridge Member of the Carmel Formation). The lower beds, formed on sand and salt flats in the Jurassic, display extreme contortion and slumping. At one time thin beds of salt (not to be confused with the much older and thicker

Pennsylvanian salt deposits) were present in these horizons, but their removal by groundwater solutions, coupled with the weight from overlying beds, caused the distorted bedding to form. These beds retarded the downward flow of groundwater before the rocks were exposed and the water that pooled in the overlying sandstone weakened the cement of the sand grains. Then, when the whole package was exposed by erosion, the overlying sandstone was more easily weathered away where the underground pockets of water had once attacked the cementing agent.

Canyonlands National Park and Dead Horse Point State Park, Utah

Nowhere are the products of fluvial erosion more superbly displayed than in the Canyonlands region of southeast Utah. Here the Colorado and Green rivers join at the great confluence to divide the landscape into three equally enchanting, but separated parts of the national park that enclose most of the scenery. Downcutting by the two rivers has delineated the Needles, Maze, and Island in the Sky districts, all lying naked under the brilliant sun with hardly a tree to mar the view. Writer Edward Abbey once wrote that "in all of this vast well of space enclosed by mesas and plateaus, nothing moves, nothing stirs. The silence is complete."

Naked red rocks ranging in age from Pennsylvanian to Jurassic are exposed along the Colorado River below Dead Horse Point, Canyonlands National Park, Utah. It was from these cliffs that the famous last ride of Thelma and Louise (of Hollywood movie fame) took place.

The Needles section, lying east of the Colorado River, is the most easily accessed section of the park and is reached by way of the paved but winding course of Utah Highway 211. Descending from the high rim of the canyons, it enters an area rich in geologic treasure. The Six Shooter Peaks display colorful aprons of the Triassic Chinle Formation capped with thin spires of Wingate Sandstone, resembling the barrel of an old pistol. The Needles District derives its name from the towerlike spires of Cedar Mesa Sandstone that dominate the landscape. The strata here record an ancient battle between sand dunes and mountain-born rivers. The reddish-brown Cutler Formation to the east is composed of gravel-rich river deposits that originated in the Ancestral Rocky Mountains. But in the Needles, the same interval consists of Cedar Mesa Sandstone, derived from the desert dunes that stopped those ancient rivers in their tracks.

The Maze, that portion of the park west of the Green River, has no paved roads and is very little visited or seen; access is difficult, but the effort is rewarded. Here lie the Orange Cliffs, an imposing land of buttes and mesas that are carved into the cliff-forming Wingate Sandstone. Between these cliffs and the river, the landscape is torn into shreds, comprising a maze of intricate canyons, and narrow-necked ridges. This section of the park features the Land of Standing Rocks and the Doll House, erosional remnants cut into the Cedar Mesa Sandstone and the overlying Organ Rock Formation.

Above: The colorful strata at the Six-Shooter Peaks in the Needles District, Canyonlands National Park, Utah

Below: Rainbow Bridge, reflected in the waters of Lake Powell, was carved into the Jurassic Navajo Sandstone near Glen Canyon National Recreation Area.

Lying north of these two sections, in the triangle between the two rivers, is the Island in the Sky. Here a paved road follows a narrow neck of land on the rim of the canyonlands, terminating at a fantastic buttress called Grandview Point. From here one can gaze into an unreal landscape of canyons and rocky ridges, framed by the often snowcapped La Sal Mountains to the east and the Abajo Mountains to the south. The spectacular Shafer Trail descends the rim and provides backcountry access to the canyons below for bikes, hikers, and four-wheel-drive vehicles. Dead Horse Point, adjacent to Island in the Sky, provides what may be the greatest single viewpoint on the entire Colorado Plateau. From the viewpoint, some fairly amazing "rock magic" is exposed. The White Rim Sandstone forms a prominent bench of the same name; however, the same sandstone is absent just a short distance to the east across the Colorado River—it has pinched out entirely across the field of view. This is the kind of paleolandscape detail that geologists dream about; the lateral changes in rock type reveal the many changing environments from the past. Geology heaven, for sure!

Glen Canyon National Recreation Area and Lake Powell, Arizona-Utah

Carved by the Colorado River, Glen Canyon exposes one of the world's thickest piles of eolian-derived sandstone. Seven different eolian units of Permian and Jurassic ages are exposed at various places in the 200-mile-long (320-km) canyon, now partly filled by the waters of Lake Powell. The upper part of the lake is crossed by Utah Highway 95 on three steel arch bridges where two Permian sandstones are exposed. The Cedar Mesa Sandstone crops out at lake level, while a thin wedge of the younger White Rim Sandstone overlies the contrasting reddish-brown sandstone and mudstone of the fluvial Organ Rock Formation. To the west, the Henry Mountains form the skyline. Southward, Glen Canyon's myriad shapes and side canyons abound. Less resistant fluvial units such as the Triassic Moenkopi and Chinle formations and the coastal plain deposits of the Jurassic Carmel Formation cause the canyon to be wider and more open. Cliffs of eolian Wingate, Navajo, Page, and Entrada sandstone, as well as the fluvial Kayenta Formation, form a narrower canyon with towering cliff walls. South of the junction with the San Juan River is a side canyon that takes you to the world's largest natural stone span, Rainbow Bridge, at almost 300 feet (90 m) tall The rounded 10,000-foot-high (3,048-m) mountain to the east is Navajo Mountain, formed as a laccolith whose sedimentary cap still conceals the intrusion (the only such occurrence on the plateau). The side canyons and slickrock terrain surrounding Lake Powell are incredible places for wilderness hiking, and slot canyons abound. Most recreational boaters do not venture far from the shoreline, leaving the backcountry for the few who choose to explore its geologic wonders.

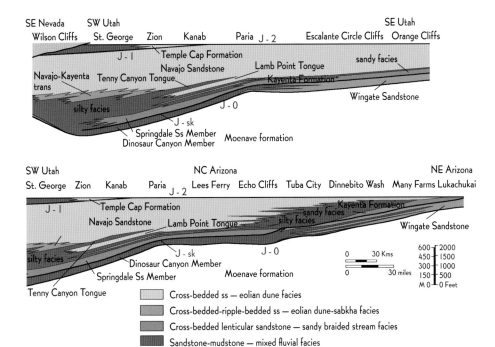

SE Nevada — SW Utah
Wilson Cliffs — St. George — Zion — Kanab — Paria — J-2 — Escalante Circle Cliffs — Orange Cliffs — SE Utah

J-1
Temple Cap Formation
Navajo Sandstone — Lamb Point Tongue — sandy facies
Navajo-Kayenta trans — Tenny Canyon Tongue — Kayenta Formation
silty facies — J-0 — Wingate Sandstone
J-sk
Springdale Ss Member — Moenave formation
Dinosaur Canyon Member

SW Utah — NC Arizona — NE Arizona
St. George — Zion — Kanab — Paria — J-2 — Lees Ferry — Echo Cliffs — Tuba City — Dinnebito Wash — Many Farms — Lukachukai

J-1
Temple Cap Formation — Kayenta Formation
Navajo Sandstone — Lamb Point Tongue — sandy facies — silty facies — Wingate Sandstone
silty facies — J-sk — J-0
Dinosaur Canyon Member
Springdale Ss Member — Moenave formation
Tenny Canyon Tongue

0 — 30 Kms
0 — 30 miles

600 — 2000
450 — 1500
300 — 1000
150 — 500
M 0 — 0 Feet

Cross-bedded ss — eolian dune facies
Cross-bedded-ripple-bedded ss — eolian dune-sabkha facies
Cross-bedded lenticular sandstone — sandy braided stream facies
Sandstone-mudstone — mixed fluvial facies

Jurassic deposits undergo many changes in the Lake Powell region. Note how the Wingate Sandstone is present at the base of both columns in the east (right) but undergoes a facies change near the Echo Cliffs and Paria River, changing into the Moenave Formation. Also, the contact with the Kayenta and Navajo formations is variable across the cross sections.

Grand Staircase–Escalante National Monument, Utah

One of the newest and most special national monuments on the Colorado Plateau is Grand Staircase–Escalante, created by presidential proclamation in September 1996. This sprawling preserve, containing 1.9 million acres (760,000 hectares), has always been well-liked by geologists, even before the days of its colorful moniker. Now, however, it is an increasingly popular destination for all kinds of backcountry enthusiasts. Rocks within the monument range in age from the Permian Kaibab Formation to the Paleogene Claron Formation. Because the monument is so extensive, it would be impossible to describe all parts of it here, so we highlight only one special area within it. One could take weeks to fully explore the many treasures of this gem.

Take U.S. Highway 89 west from Page and stop first at the Big Water Visitor Center. Here is a small but wonderful paleontology exhibit. You can learn about and easily visualize aspects of the Cretaceous landscape of the plateau. Bones of a giant plesiosaur and colorful murals depict scenes of the Cretaceous Interior Seaway. You can check here for current road conditions within the monument (most are dirt routes). One of the best is the Cottonwood Canyon Road, a bumpy forty-five miles (72 km) of upturned rock bliss that follows the East Kaibab monocline. Don't miss a trip on this road if you are headed to Bryce, but don't attempt it if it has rained recently. It is often passable to passenger vehicles and is a rewarding trip that displays fantastic erosion processes on the monocline. Traveling north on the road is probably the most scenic direction. An easy one-mile (1.6-km) side road will take you to scenic Grosvenor Arch. Look for oyster shells embedded in the Dakota Sandstone here.

Beautiful display of the Glen Canyon Group along the Escalante River (lower right) near Grand Staircase–Escalante National Monument, Utah

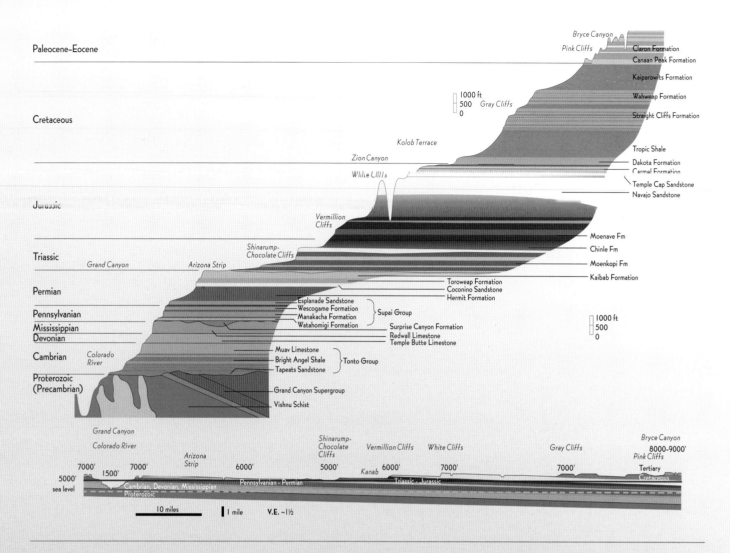

If the Cottonwood Canyon Road is impassable, stay on the pavement of Highway 89, and you'll soon pass the White House Trailhead, the beginning of a spectacular hike into some of the deepest slot canyons on the plateau. A four- to six-day hike takes you to the mouth of the Paria River at Lees Ferry, Arizona.

Beyond the Paria River, Highway 89 cuts through the upturned trend of the East Kaibab monocline. This is the major structural feature that bisects the monument. It begins far to the south of Grand Canyon but extends north for more than one hundred miles (160 km). The monocline marks the transition from the Glen Canyon area to the east to the Grand Staircase to the west. One of the best places for geologists to visit is the little-known valley surrounding the site of the oldest settlement in southern Utah, Old Paria. Little is left of the original town, but the movie set built in the 1960s hangs on, even though flash floods and tourists attempt to erase it. As always, the geology abides.

Exploration of this geologic gem begins on a dirt road that leaves Highway 89 and heads north. The road is built upon a platform of Permian Kaibab Formation. Proceeding northeast and downhill, one actually goes up-section into the younger

Triassic rocks of the Moenkopi and Chinle formations. This is because the rocks dip more steeply on the monocline than the road descends on the topography. Just before plunging into the valley and canyon of the Paria River, you are rewarded with one of the Colorado Plateau's most magnificent views. The monocline lies before you at Gingham Skirt Butte, and the Paria River has carved a narrow canyon behind it. The brightly tinted sandstones and mudstones of the Triassic and Jurassic formations are framed in a backdrop of the somber grays and tans of the marine Cretaceous rocks of the Kaiparowits Plateau. To the west, the Vermilion and White Cliffs stretch to the distant horizon, delineating the central heights of the Grand Staircase. The scene is marvelous and full of color. The Grand Staircase was first noted by the distinguished geologist, Clarence Dutton. It describes a section of rocks that are stacked in colorful cliffs from the bottom of the Grand Canyon to the top of Bryce Canyon.

Grosvenor Arch, Grand Staircase–Escalante National Monument, Utah

Bryce Canyon National Park, Utah

Bryce is truly different. Here the pink, orange, and white limestone and conglomerate of the Claron Formation (formed 40 to 50 million years ago) are whittled by repeated freeze-thaw cycles into delicate hoodoos and spires. These sentinels soar skyward as if in competition with the scented, green conifers that blanket its rim. The park's elevations range from 8,000 to 9,000 feet (2,400–2,700 m) above sea level. Bryce and neighboring Cedar Breaks National Monument (at 10,000 feet/3,048 m) form a colorful top to the Grand Staircase and are known as the Pink Cliffs. The Claron Formation (called Wasatch Formation in the older literature) formed in a Laramide-age basin between adjacent uplifts. Most of the conglomerate was carried by rivers that drained from the Mogollon and Sevier highlands far to the south and the west. This means that the Grand Canyon was not a barrier to fluvial transport at the time. Explosive volcanism covered the area north of the park about 25 to 30 million years ago.

Although the rim of Bryce Canyon rises above the Paria Amphitheater, the escarpment is actually carved into rocks on the down-thrown side of the Paunsaugunt fault. The reason for this is that the down-thrown Claron rocks are harder and more resistant to erosion than the rocks on the up-thrown side of the fault, which consist of very soft Cretaceous shales. The Paria River and its tributaries have eroded in a headward fashion up the fault trend and nibbled away much faster and more readily into these soft rocks. The rate of retreat at Bryce, nonetheless, has been determined to be a whopping one foot (0.3 m) every sixty-five years—about ten times the estimated erosion rate of the Mogollon Rim near Sedona, Arizona. This means that the retreating cliff face at Bryce will erode west into the bed of the East Fork of the Sevier River in about 600,000 years. Stream capture of the Sevier by the Paria River points out how the larger Colorado River may have developed, for when

The Claron Formation, representing the top of the Grand Staircase, mantled in snow at Bryce Canyon National Park, Utah.

larger rivers extend their upstream reach and intercept smaller rivers, new tracts of drainage are added to the more vigorous and expansive rivers. Slowly but steadily, the shaping of Bryce's pinnacles keeps the rock faces sharp and fresh. The view at the south end of the escarpment from Yovimpa Point looks down across the area of the Grand Staircase and is a scene of beauty never to be forgotten.

Zion National Park, Utah

Zion National Park is one of Earth's very special places. It is one of the few canyon systems on the Colorado Plateau where the major access is from the canyon floor and the views are upward into the cliffs. This one difference makes Zion a worthwhile stop for any geologic explorer. Many trails lead up to pools, waterfalls, arches, and vistas. Zion Canyon proper is easily accessed by taking the free shuttle to the many different stops that allow a close look at the geology. Weeping Rock is a spring that emerges from the Kayenta-Navajo contact. A walk up the Zion Canyon Narrows is a cool treat on a hot summer day. A paved highway leads east through the 1.1-mile-long (1.8-km) Zion Tunnel and a wonderland of cross-bedded Navajo Sandstone outcrops. Checkerboard Mesa displays this eolian deposit rather well.

Sunset view of the West Temple, Zion National Park, Utah. The lower slopes (partly in shadow) are cut into the Triassic Moenkopi Formation. The Shinarump Member of the Chinle Formation forms the bench (center) and the upper cliff (top) is composed of the Jurassic Navajo Sandstone.

The Mesozoic section of sedimentary rocks found here is second to none and the scenery truly inspires awe. Southwest of the entrance in the Virgin River valley, colorful badlands of the Moenkopi Formation expose myriad Triassic coastal plain, tidal flat, and fluvial environments. The prominent yellow-brown cliff of the Shinarump Member of the Chinle Formation marks a transition to more vigorous streams that blanketed the area. The overlying Petrified Forest Member documents sluggish meandering streams that covered much of the Southwest during the Late Triassic. At the south entrance to the canyon the Jurassic Moenave Formation is well exposed above the visitor center and Watchman Campground. This formation was deposited in both ephemeral and perennial braided streams.

The red, slope-forming fluvial Kayenta Formation rises steeply to the foot of the earth's most impressive cliffs, the 3,000-foot-high (900-m) walls of Zion Canyon. The walls expose a vast eolian dune system deposited grain by grain, layer by layer, to form the Navajo Sandstone. Cross-beds, some nearly one hundred feet (30 m) high, mark the passage of the ancient dunes. Another desert dune system deposited the overlying Temple Cap Sandstone. On top of the cliffs, out of view from the canyon floor, are the marine and shoreline deposits of the Carmel Formation. Cretaceous rocks, almost two miles (3.2 km) thick in some areas, directly overlie the Carmel and are exposed along portions of the road between Zion and Bryce. Some of the Jurassic rocks were removed by erosion in the Nevadan Orogeny. Zion Canyon and its side canyons and alcoves are best explored on foot, and numerous trails access these incredible places.

A more remote part of Zion is the Kolob Canyons section of the park located just east of Interstate 15 between Cedar City and Saint George. Spectacular exposures of Navajo Sandstone and the Kayenta, Moenave, and Chinle formations can be found along the six-mile (9.6-km) paved road or the few trails that lead away from it. Also, do not miss the Dinosaur Discovery Site at Johnson Farm in nearby Saint George, where you can see a spectacular set of dinosaur footprints in the Kayenta Formation and enjoy the wonderful interpretive display.

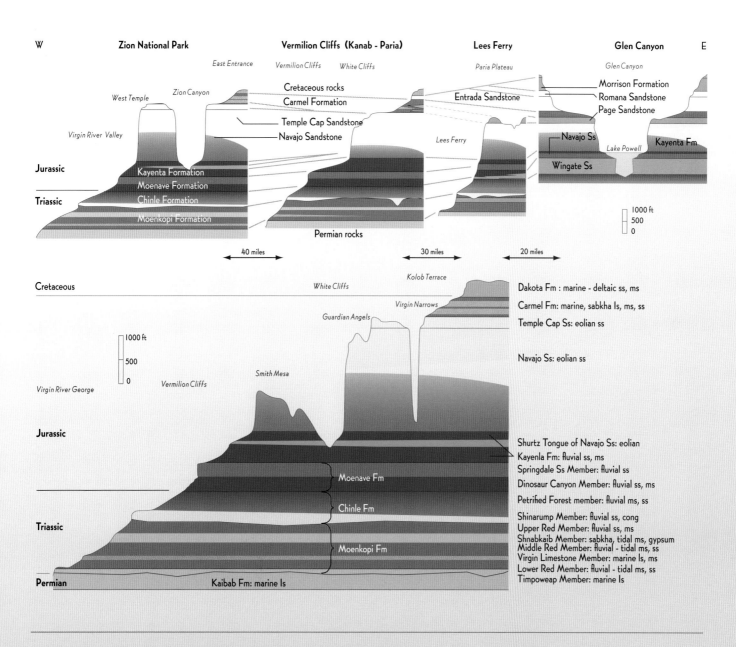

Correlation chart and cross sections showing the surface and subsurface relationships of Mesozoic rocks from Zion (left) to Glen Canyon (right).

Grand Canyon National Park, Arizona

The Grand Canyon is the earth's most stupendous erosional landform, and it may be the earth's largest canyon. Measured along the Colorado River, the Grand Canyon is 277 miles (443 km) long, 10 to 18 miles (16–29 km) wide, and up to 6,000 feet (1,800 m) deep. Within its walls, 1,750 million years of earth history is exposed in dizzying detail. The parade of geologic events represented in this area is nearly unparalleled. Grand Canyon is a fitting place to end this survey of the choicest geologic areas on the Colorado Plateau.

The Inner Gorge of Grand Canyon exposes Precambrian rocks that document how this part of the North American continent was created and eventually attached to a larger nucleus of North America. On top of these crystalline rocks, younger Precambrian sediments document long periods of deposition, mountain building, and erosion—all before the evolution of advanced forms of life. The overlying 4,000 feet (1,200 m) of Paleozoic section that form 80 percent of the upper canyon walls represent one of the most continuous and well-exposed outcrops in North America and perhaps the world. The Cambrian, Devonian, Mississippian, Pennsylvanian, and Permian rocks document in great detail the depositional history during those periods of time. Mesozoic rocks are found to the west, north, and east of the park, and their outcrop pattern suggests that they, too, once covered the Grand Canyon before erosion stripped them away. The Cenozoic uplift history is read from the numerous folds and faults that disrupt the sediments. The Butte fault, named by the eminent geologist C. D. Walcott in the 1890s, and the East Kaibab monocline, both in eastern Grand Canyon, are perhaps the most well known of these structures.

The canyon was cut during the Cenozoic, although there remains considerable debate as to exactly how and when the Colorado River carved the canyon. The most widespread theories suggest a fairly young age for the canyon at about 6 million years. Other evidence indicates that some parts of the canyon (both in area and depth) may have been cut much earlier than this. This suggests that completely separate river systems may have become integrated at some point in time into the river system we see today. This may have been accomplished by headward erosion, the catastrophic overflow of ancient lakes, or the collapse of subterranean karst beneath the Kaibab upwarp. Regardless of the age of the canyon, there is no debate that the Colorado River did it. Most geologists believe that the river has always been approximately as wide as it is today (about 300 feet/100 m). The river itself, therefore, could only have cut that much of it. Other forces of erosion such as freezing and thawing, side drainage enlargement, and gravity have widened the canyon significantly. These processes continue today.

The Little Colorado River enters the Grand Canyon and exposes 4,000 feet (1,200 m) of Paleozoic strata from the Tapeats Sandstone (far right) to the Kaibab Formation (top).

The best way to study and experience Grand Canyon is on foot along one of the maintained trails along the rim or leading into the canyon. An overnight loop hike down the South Kaibab Trail to Phantom Ranch and up the Bright Angel Trail is a geologic classic. (Total distance round-trip is sixteen and a half miles/27 km. A backcountry permit is required if you camp overnight.) Viewpoints on both the North and South rims provide a means to see the broad relations between various rock units, to appreciate the vast power of erosional processes, and to observe the overall sequence of events that created the vast landscape.

Alternatively, if one has the time and resources, a raft trip through the canyon on the Colorado River is perhaps the best way to view the entire geologic profile up close. This is perhaps one of the most amazing trips that anyone can undertake, and the much-heralded rapids are adequately spaced and need not be a deal breaker. This option allows one to view all of the canyon's rocks, including the recent lava flows in the western part of the canyon that repeatedly dammed the river in spectacular fashion, beginning 630,000 years ago. The Grand Canyon truly does have it all!

The Grand Canyon of the Colorado River

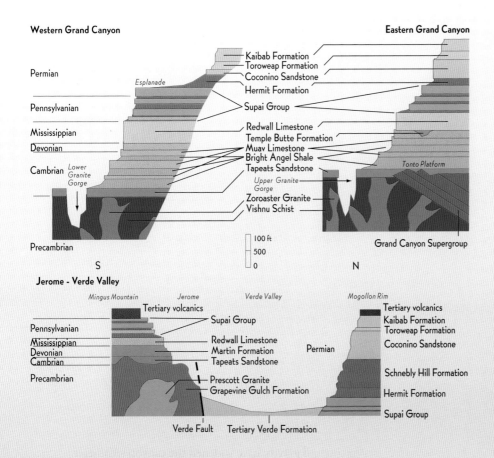

Stratigraphic relationships of Paleozoic rocks between eastern and western Grand Canyon (top), and from Mingus Mountain to the Mogollon Rim in the Verde Valley (bottom).

Appendix
Paleogeography, Paleogeographic Maps, and How They Are Made

The word paleo is of Greek origin and means "old" or "early." Literally then, paleogeographic maps are depictions of an old or previous geography. They show the location of ancient seas, shorelines, swamps, dunes, rivers, beaches, plains, deserts, or mountains. The paleogeographic maps presented here are some of the most intricate and detailed ever published, whether in professional or popular literature. The maps portray the dynamic landscape changes that occurred on the Colorado Plateau, in North America, and around the world throughout 1,750 million years of earth history. But what is the true value in knowing ancient geography? What benefit do these maps provide the professional and the curious layperson? How are the maps made?

First and foremost, paleogeographic maps are the ultimate synthesis of innumerable regional geologic studies. Ever since the first American explorers entered the West in the 1840s, the geologists that accompanied them have sought to understand the origins of this unique landscape. Since that time the number of studies has grown exponentially. This has created a body of knowledge that is incredibly rich in its separate parts, but oftentimes lacks synthesis of those disparate parts. These paleogeographic maps take myriad geologic information, garnered through multiple decades of research, and blend it into a coherent whole which, in and of itself, can be regarded as original research. The amount of work involved in sorting and assembling the data and constructing the maps is difficult to comprehend.

One of the most important concerns in constructing paleogeographic maps is determining how widely separated and distinctly different rock units may correlate with one another. This may be difficult since rock layers are often eroded away, pinched out, or are still buried and not directly observable. Most widely separated rock units on the plateau can be correlated simply by becoming familiar with the specific sequence of rock strata. In many instances this does not change appreciably from one place to the next. If formations do change their appearance slightly from place to place, they will still be set between similar unchanged strata, or may contain recognizable fossils that allow for a correlation. Volcanic ash beds, which can be accurately dated in sedimentary strata, are another excellent correlation tool.

However, some formations pinch out, drastically change their rock type, or have no fossils or ash beds within them. In these instances the correlations between rock strata are much more difficult to establish and raging controversies have sometimes ensued between geologists who couldn't see eye to eye. Today, the relationships between rock units on the Colorado Plateau are well known and remain controversial in only a handful of instances and of such limited scope as to not hinder their overall association. Usually it is possible to refine conflicting data, and most geologists can agree on what the rocks represent. Once this is established, the size and placement of an ancient seaway can be depicted on a map by noting the extent of the limestone outcrops from which the evidence was collected. Likewise, the extent of exposed mudstone can determine the presence of an ancient river system feeding that sea.

After organizing vast amounts of scattered data, the geologist is ready to construct a map. Computer paint-and-fill techniques are employed to create specific images for the map. Sometimes actual photographs from space of, say, the delta of the Ganges River or a Sahara Desert dune field, are electronically copied or cloned. All of these data are then scanned into a pixel-oriented computer program. Depictions of these landscape elements can then be skillfully stretched, skewed, rotated, resized, repeated, cloned, or blended. Space photographs of a Sahara dune field can be used to show part of the paleogeography of the Navajo Sandstone. In this way, actual images of modern dunes, rivers, and deltas realistically depict the postulated position of these ancient landscapes. The process begins anew for each map, ultimately revealing how the landscape evolved through time.

These ancient environments are plotted on a base map that contains known reference points such as county and state lines or international boundaries. These reference points allow the viewer to recognize where a landscape element was specifically located in the modern landscape. Separate layers of information are created that relate this paleogeography to other data or text. Examples of such layers might be depictions of the ancient lines of latitude and longitude, including the equator and the poles. This information can be superimposed on the ancient landscape elements that gave rise to the deposits. These layers, each created separately, can be added or removed as desired. The simplest views show only the ancient geography without any overlying text or other data. In this book political boundaries are always included as reference points to orient the viewer. These lines are layered in very light tones, however, in order not to detract from the artistry of the maps. Since books are necessarily static with respect to adding or subtracting layers, and knowing that our foremost goal was to spotlight the beauty of the ancient geography, we present the largest treatment of each map without any text-rich layers.

A set of paleogeographic maps can show the spatial and temporal distribution of ancient landscapes, but they should be considered hypothetical in a certain sense. On average, formation of a single rock unit may occur over a period of 5 to 10 million years. Yet this vast amount of time is shown in only a few snapshots which approximate the geography of that period. Furthermore, rock formations today are either partially eroded away or still deeply buried. Thus, they cannot divulge all of the spatial details that the maps appear to represent. Nonetheless, thumbing through these pages will produce an acceptable sense of this change, especially for those time periods that preserve the greatest amount of detail: Pennsylvanian, Permian, Triassic, Jurassic, and Cretaceous.

Maps showing the oldest geography are the most generalized. The Precambrian, Ordovician, and Silurian time periods preserve the least evidence; consequently, those maps are the most hypothetical. Maps for the Pennsylvanian through Cretaceous time periods are more precise because those rock units are more widely exposed on the plateau and have the fewest gaps in the rock record. Better and more complete exposures mean better data, better interpretations, and more accurate maps. However, keep in mind that even though the extent of individual dune fields and the location of specific streams can be documented on the ground, a reconstruction is still hypothetical because parts of that same deposit may be removed by erosion or still buried beneath younger rocks elsewhere. No matter how detailed the study, it is impossible to reconstruct every detail from the geologic past. This is especially true for depictions that attempt to show the specific location of ancient river channels.

Where does the data that is used to reconstruct these maps originate? All of it comes from the rocks themselves, whether exposed on the surface or penetrated by water and oil wells. Geologists look at the incredible details preserved in rocks to make interpretations. These include fossils, mineral content, geochemical signatures, and the size, shape, texture, and orientation of individual mineral grains. Fossils are the remains of the ancient life-forms that existed at the time the rock was deposited. They yield clues to the specific environments in which the rocks formed and can also be used to correlate widely separated rocks of the same age. The mineral content in a sedimentary rock can point to the direction from which the sediment came. Some minerals do not travel far in rivers; others are quite durable and make it all the way to the sea. Grains of sediment are always arranged in particular ways and give important clues that indicate whether the rock's sediment was deposited in rivers, dunes, seas, or shorelines. The arrangement of rock layers in time (the order in which they are vertically and sequentially stacked) and space (the extent to which they spread across the landscape) also provides evidence of their origin.

Reading the maps is relatively straightforward. The hills, mountains, rivers, dunes, plains, and coastlines are obvious to anyone who studies maps. Snow on the mountains implies height but does not necessarily depict a glacier, and mountains shown without snow may have been high enough for occasional or seasonal snow. Conical peaks represent volcanoes and may be shown with craters on their summits. Solid lines depict perennial streams and dashed lines imply ephemeral systems. Braided and meandering rivers are also shown. As mentioned before, however, the specific placement of rivers and streams is mostly generalized. The colors on land reflect several things including the postulated climate. Maps from the Devonian to the present use browns, reds, and tans to imply times of arid conditions, while shades of green imply humid conditions. (Land plants were absent in the geologic record before the Silurian, therefore colors on the Precambrian and Cambrian maps denote landscapes without vegetation.) Yellow and orange colors that show wavy textures reflect dune deposits in desert settings.

In marine environments, the sediment types are broadly reflected as follows: soft yellow depicts sandy bottoms, grays and tans show muddy deposition, and clear pale to medium blues show carbonate deposition. Water depth in the seas and oceans is shown on a relative color scale with deep blue representing oceanic (abyssal) depths and light blue showing shallow marine conditions. Submarine canyons are also shown on the shelf edges. Reefs are shown as white stipples in the water in either shallow water or at the edge of a continental shelf. Blue-black lines in the oceans are deep trenches, and ridges are offset with narrow dark lines that represent spreading centers or mid-ocean ridges. Using all of these color-coded techniques, which are used on most modern physical maps, these ancient landscapes can be depicted in realistic fashion.

Below: Paleogeographic maps can be constructed to show what future landscapes might look like. This Mollweide (oval) projection, based on present tectonic patterns and rates, suggests what earth may look like 100 million years into the future. Note how Australia has drifted into and collided with China; East Africa has been rifted away from mainland Africa; the Persian Gulf has closed; and the Atlantic Ocean has been much wider.

GLOSSARY

accommodation space: Space in which sediment is deposited and preserved; space created by some combination of subsidence and rise in base level, usually sea level

active margin: Edge of a continent bordered by plate margin; usually a zone of active tectonics; *cf.* passive margin

alluvial (alluvium): General term applied to all deposits formed by running water

alluvial fan: L-shaped landform and deposit at a point where steep-gradient (mountain) streams become shallow-gradient and deposit sediment

arch, upwarp, uplift: Area that is uplifted relative to surrounding areas of subsidence; *cf.* basin

basement: Foundation of continent composed mainly of igneous and metamorphic rocks, commonly covered by thin veneer of sedimentary rocks

basin: Area that is low or down-dropped relative to surrounding areas of uplift; *cf.* uplift

Basin and Range Province: Geologic province west of Colorado Plateau that was subjected to Neogene faulting and consists of alternating mountain ranges and intervening basins

batholith: Large mass of plutonic (intrusive) igneous rock that formed within the earth's crust; exposures of batholiths imply significant uplift, causing rock to surface

bedform: Any positive feature in noncohesive sediment formed by wind or water currents, including ripples and dunes. Migration of bedforms produces cross-bedding.

carbonate (rock or sediment): Collective term of any rock or sediment composed of $CaCO_3$ or $[Ca, Mg]CO_3$; includes limestone and dolomite

chert: Fine-grained quartz, commonly occurs as nodules in sedimentary rocks that can be later reworked as clasts into younger sediments and sedimentary rocks

clast: Any particle weathered from preexisting rock and included in sediment and sedimentary rock

coastal plain: General setting of low, flat terrain adjacent to coasts

coeval: Of or approximately the same time

Cordilleran Arc: Arc and subduction complex on west margin of the Americas; related to subduction of Pacific Ocean plates from Triassic to present

core: Central layer of Earth

correlation (of rock units): Determination of age relations (time-rock stratigraphic) of rocks from one place to another place

craton: The tectonically stable, generally central part of a continent; usually underlain by Precambrian igneous and metamorphic rocks

crust: Thin outermost layer of Earth

diapir: Structure that flows or intrudes into another body of rock, especially salt

dune: Large (> 8 inches/30 cm high) sandy bedforms created by wind and running water

eolian, aeolian: Environment or process of the wind, depositional setting of windblown dunes

environment (depositional): Site where sediment is deposited; sum of all physical, chemical, and biologic processes at the site that act on sediment and produce sediment's characteristics

ephemeral: Active only part of the time, usually applied to streams that flow infrequently or intermittently

epicontinental sea: Sea upon the continent, especially interior seas; Hudson Bay and North Sea are modern examples.

erg: Large areas of windblown sand (Arabic); sand seas

estuarine, estuary: Of or pertaining to wide river mouths where marine processes penetrate into river mouth; used to describe setting, processes, and products

eustacy: Of or pertaining to global sea level; changes in sea level of global, rather than local effect

evaporate: Salt deposits formed by evaporation; gypsum, halite, and potash are common examples.

extension: Tectonic setting in which crust is being pulled apart

fluvial: General term referring to rivers (streams) and their deposits

foliation: Pattern in metamorphic rocks whereby minerals align parallel to each other; allows geologists to infer direction of forces that produced metamorphism

forearc basin: Basin located between trench and volcanic arc in subduction complex

foreland basin (foredeep): Basin formed adjacent to stacked thrust sheets; thrust sheets bend adjacent crust downward causing subsidence

formation: Formal rock-stratigraphic term; any mappable rock body that can be separated from any adjacent rock body; in areas where formal rock stratigraphy has been established, all rocks must be assigned to one or more formations

Ga: Giga-annum or billions of years before present

group: Formal rock stratigraphic term; consists of two or more formations thought to be closely related in origin

hingeline: Linear zone that marks sharp change in subsidence rate of sedimentary basin; separates thin strata from thick strata

intertonguing (of rock layers): Alternation of contrasting lithologies where one lithology pinches out into contrasting lithology; caused by shifting back and forth of depositional settings with contrasting lithology

island arc: Line of volcanoes, usually arcuate in map view, that mark location of deep-crustal melting associated with subduction

karst: Type of landscape developed on extensive limestone terrain characterized by solution features such as sinkholes, caves, and disappearing streams

laccolith: Pancake-like igneous intrusion that domes up or pushes apart sedimentary rocks that it intrudes; usually fed by central plug or stock

lacustrine: General term referring to lakes and their deposits

laminae: The finest visible layering or stratification in a sedimentary rock; "paper thin"

lithified: Cemented; to become a rock; hardened

lithology: Literally, rock type; common field terms include sandstone, limestone, mudstone, granite, schist, etc.

Ma: Mega-annum or millions of years before present

magma: Molten, mobile rock material within crust of Earth; can become intrusive or extrusive

mantle: Interior layer of Earth between core and crust; composed of ultramafic silicate rock.

marine: Of or related to the sea; environment characterized by normal marine salinity of 3.5 percent

member: Subdivision of a formation

mass wasting: A general term for a variety of processes by which large masses of earth material are moved by gravity, either quickly or slowly, from one place to another

microcontinent: General term applied to small block or part of a plate that consists of continental crust; a continental block of small proportions

monocline: Fold typical of Colorado Plateau with a single steep limb; in theory a steplike bend that separates uplifted and down-dropped horizontal strata, but more commonly the steep flank of an asymmetrical anticline-syncline pair

monsoon: Climate pattern in which summer months generate more precipitation than winter months; Southeast Asia is typical example

orogenic, orogeny: Of or related to mountain building

paleocurrent: Some indicator in sedimentary rock such as cross-bedding that yields ancient flow direction

paleogeograhy: Literally "ancient geography"; refers to how Earth looked in the past

Pangaea: Supercontinent that existed from Pennsylvanian through Jurassic that contained 80–90 percent of all continental blocks assembled into one continent

passive margin: Edge of continent and adjacent ocean on same plate; zone of little or no tectonic activity; *cf.* active margin

perennial: Active all of the time; usually applied to rivers (streams) that flow year-round

period: Formal subdivision of geologic time, usually with a duration from 30 to 60 million years

Permo-Triassic: Shorthand for Permian and Triassic

plates (tectonic): Rigid piece of lithosphere (uppermost mantle and overlying crust) that moves as a unit and interacts with other plates at active plate boundaries; most present plates contain both continental and ocean crust in various proportions

point bar: Prominent landform or deposit of meandering streams that forms at inside bends of meanders where sediment accumulates

protolith: Parent rock, used in context of metamorphic rocks, referring to what rock type was before metamorphism

radiometric dating: Measurement of ratio of parent and daughter isotopes in unstable isotope system that allows determination of absolute age of rock or mineral

regression: Process-event in which sea withdraws from area of land; *cf.* transgression

rift/rifting: Splitting or pulling apart of continents or ocean basins so that extension of crust occurs; usually accompanied by normal faulting; extensive rifting results in new plate formation wherein a rift separates one plate into two.

Rodinia: Supercontinent present in Late Precambrian that contained almost all continental material of the time

sabkha: Of or pertaining to a place or environment and its processes in an arid setting with a high water table; they can be coastal or inland; modern sabkhas (spelled numerous ways) are common in Africa, Australia, and Arabia

sedimentary basin: A low area where sediments accumulate and are preserved; because basins subside through geologic time, they are characterized by thick sedimentary deposits relative to surrounding regions

series (geologic time): A rock-stratigraphic term that is a formal subdivision of a system

shelf, platform: A broad, low area generally at or slightly below sea level; the site of broad, thin (compared to basins) sedimentary deposits of marine and shoreline origin

siliciclastic (rock or sediment): A clastic deposit (grains can be any size) containing silica-bearing minerals, most commonly quartz, feldspar, or clay; *cf.* carbonate

stratigraphic, stratigraphy: Of or pertaining to the study and classification of stratified rocks

subduction: Process at active plate margin where one plate dives (or is pulled) beneath another plate; occurs at convergent margins

subsidence: Sinking of an area through geologic time; opposite of uplift or mountain building

supercontinent: Assembly of several continents by collisional orogenic events; commonly contain all or most of continents that exist at given time; Pangaea and Rodinia are examples

Supergroup: Formal rock stratigraphic term; consists of two or more groups of rocks thought to be closely related in time or origin

suture: Zone of joining or contact between continents that have collided; most orogenic zones have formed by this process. Most sutures are marked by ophiolites, squeezed and commonly metamorphosed pieces of old ocean crust that lay between collided continents within suture zone

system (time-rock): Time-rock equivalent of period; system is used when referring to rocks of a given time whereas period is used when only time is being discussed. Time occurred whether or not rocks were formed to record geologic history, hence, the reason to differentiate the two terms.

tectonic: Of or pertaining to large-scale to global structure and structural processes; especially applied to study of large-scale structures through geologic time

terrestrial, terrigenous: Of or pertaining to the land or the continents, especially the sediment derived from these areas

time transgressive: Refers to geologic process that requires amounts of geologic time to move across a region; mountain building (orogeny) and transgression-regression are examples

trace fossils: General term applied to any fossils other than body parts: tracks, trails, burrows, resting marks

transform boundary: A plate boundary where the stresses are horizontal rather than vertical

transgression: Process-event in which sea covers or moves across what was once land; *cf.* regression

ultramafic ophiolites: An assemblage of igneous rocks such as basalt, gabbro, and peridotite that are rich in dark minerals and that form near the oceans' spreading centers

unconformity: A surface or bedding plain in layered rocks that represents significant geologic time without a rock-sediment record, evidence missing from rock record

uniformitarianism: A geologic doctrine which states that processes and physical laws have a consistency or uniformity over large amounts of time; sometimes misused to state that conditions remain the same through geologic time, which is incorrect. Geologists paraphrase uniformitarianism as "the present is the key to the past." Although uniformitarianism is widely upheld by modern geologists, we also know that the doctrine must be applied with restraint.

upwarp, uplift: Area that is uplifted relative to surrounding areas of subsidence; *cf.* basin

Western Interior: General region of western North America inland from the coast that usually includes the Rocky Mountains (U.S. and Canada), Colorado Plateau (U.S.), part of all of the Basin and Range Province (U.S. and Mexico), and westernmost High Plains (U.S. and Canada)

Abbott, L., and T. Cook. 2007. *Geology underfoot in northern Arizona*. Missoula: Mountain Press Publishing.

Baars, D. L. 1962. Permian system of the Colorado Plateau. *Bulletin of the American Association of Petroleum Geologists* 46:149–218.

Baars, D. L. 1995. *Navajo country*. Albuquerque: University of New Mexico Press.

Bjerrum, C. J., and R. J. Dorsey. 1995. Tectonic controls on deposition of Middle Jurassic strata in a retroarc foreland basin, Utah-Idaho trough, Western Interior, USA. *Tectonophysics* 14:962–978.

Blakey, R. C. 1974. Stratigraphic and depositional analysis of the Moenkopi Formation, southeastern Utah. *Bulletin 104*. Salt Lake City: Utah Geological and Mineral Survey.

_____. 1989. Triassic and Jurassic geology of the southern Colorado Plateau in geologic evolution of Arizona. *Arizona Geological Society Digest* 17:369–396.

_____. 1990. Stratigraphy and geologic history of Pennsylvanian and Permian rocks, Mogollon Rim region, central Arizona and vicinity. *Bulletin 102*. Boulder: Geological Society of America.

_____. 1990. Supai Group and Hermit Formation, in Beus, S. S., and M. Morales, eds. *Grand Canyon Geology*. New York: Oxford University Press.

_____. 1994. Paleogeographic and tectonic controls on some Lower and Middle Jurassic erg deposits, Colorado Plateau, in Caputo, M. V., J. A. Peterson, and J. J. Franczyk, eds. *Mesozoic systems of the Rocky Mountain region, USA*. Denver: Rocky Mountain Section, Society of Economic Paleontologists and Mineralogists (SEPM).

_____. 1996. Permian eolian deposits, sequences, and sequence boundaries, Colorado Plateau, in Sonnenfeld, M. D., and M. W. Longman, eds. *Paleozoic systems of the Rocky Mountain Region*. Denver: Rocky Mountain Section SEPM

_____. 2007. Paleogeography and geologic history of the western Ancestral Rocky Mountains, Pennsylvanian-Permian, southern Rocky Mountains and Colorado Plateau, in Houston B., P. Moreland, and L. Wray, eds. *The Paradox Basin: Recent advancements in hydrocarbon exploration*. Denver: Rocky Mountain Association of Geologists.

_____. Forthcoming. Pennsylvanian-Jurassic sedimentary basins of the Colorado Plateau and Southern Rocky Mountains, in Miall, A. D. ed. *Sedimentary Basins of North America*. Amsterdam: Elsevier.

_____, and R. Gubitosa. 1983. Late Triassic paleogeography and depositional history of the Chinle Formation, southern Utah and northern Arizona, in Reynolds, M. W., and E. D. Dolly, eds. *Mesozoic Paleogeography of the west-central United States: Rocky Mountain Paleogeography Symposium 2*. Denver: Rocky Mountain Section SEPM.

_____, F. Peterson, and G. Kocurek. 1988. Late Paleozoic and Mesozoic eolian deposits of the Western Interior of the United States. *Sedimentary Geology* 56:3–125.

_____, and R. Knepp. 1989. Pennsylvanian and Permian geology of Arizona, in Jenney, J. P., and S. J. Reynolds, eds. *Geologic evolution of Arizona*. Tucson: Arizona Geological Society.

_____, E. L. Basham, and M. J. Cook. 1993. Early and Middle Triassic paleogeography, Colorado Plateau and vicinity, in Morales, M., ed. *Aspects of Mesozoic geology and paleontology of the Colorado Plateau*. Flagstaff: Museum of Northern Arizona.

_____, K. G. Havholm, and L. S. Jones. 1996. Stratigraphic analysis of eolian interactions with marine and fluvial deposits, Middle Jurassic Page sandstone and Carmel Formation, Colorado Plateau, USA. *Journal of Sedimentary Research* 66:324–342.

Burchfiel, B. C., P. W. Lipman, and M. L. Zoback. 1992. *Decade of North American Geology, Volume G-3, The Cordilleran orogen: conterminous U.S.* Boulder: Geological Society of America.

Busby-Spera, C. J. 1988. Speculative tectonic model for the early Mesozoic arc of the Southwest Cordilleran United States. *Geology* 16:1121–1125.

Campbell, J.A. 1980. Lower Permian depositional systems and Wolfcampian paleogeography, Uncompahgre basin, eastern Utah and southwestern Colorado, in Fouch, T. D., and E. R. Magathan, eds. *Paleozoic paleogeography of the west-central United States*. Denver: Rocky Mountain Section SEPM.

Cather, S. M., S. D. Connell, R. M. Chamberlin, W. C. McIntosh, G. E. Jones, A. R. Potochnik, S. G. Lucas, and P. S. Johnson. The Chuska erg: Paleogeomorphic and paleoclimatic implications of an Oligocene sand sea on the Colorado Plateau, in *Bulletin 120*. Boulder: Geological Society of America.

Chan, M. A. 1989. Erg margin of the Permian White Rim Sandstone, SE Utah. *Sedimentology*. 36:235–251.

Clemmensen, L. B., H. Olsen, and R. C. Blakey. 1989. Erg-margin deposits in the Lower Jurassic Moenave Formation and Wingate Sandstone, southern Utah, in *Bulletin 101*. Boulder: Geological Society of America.

Crabaugh, M., and G. Kocurek. 1993. Entrada Sandstone: an example of a wet aeolian system, in Pye, K. ed. *Special Publication 72: The dynamics and environmental context of aeolian sedimentary systems*. London: Geological Society of London.

Currie, B. S. 1997. Sequence stratigraphy of nonmarine Jurassic-Cretaceous rocks, central Cordilleran foreland-basin system, in *Bulletin 109*. Boulder: Geological Society of America.

Dickinson, W. R. 1989. Tectonic setting of Arizona through geologic time, in Jenney, J. P., and S. J. Reynolds, eds. *Digest 17: Geologic evolution of Arizona*. Tucson: Arizona Geological Society.

Dickinson, W. R., and W. S. Snyder. 1978. Plate tectonics of the Laramide orogeny, in Matthews, V. ed. *Laramide folding associated with basement block faulting in the western United States, Memoir 151*. Boulder: Geological Society of America.

Doelling, H. H. 1975. *Geology and mineral resources of Garfield County, Utah: Bulletin 107*. Salt Lake City: Utah Geological Survey.

Doelling, H. H., and D. D. Fitzhugh. 1989. *The Geology of Kane County Utah: Bulletin 124*. Salt Lake City: Utah Geological Survey.

Dubiel, R. F. 1994. Triassic deposystems, paleogeography, and paleoclimate of the Western Interior, in Caputo, M. V., J. A. Peterson, and J. J. Franczyk, eds. *Mesozoic systems of the Rocky Mountain region, USA*. Denver: Rocky Mountain Section SEPM.

Dubiel, R. F., J. T. Parrish, J. M. Parrish, and S. C. Good. 1991. The Pangaean megamonsoon: Evidence from the Upper Triassic Chinle Formation, Colorado Plateau. *Palaios* 6:347–370.

Dubiel, R. F., J. E. Huntoon, S. M. Condon, and J. D. Stanesco. 1996. Permian depositional systems, paleogeography, and paleoclimate of the Paradox Basin and vicinity, in Sonnenfeld, M. D., and M. W. Longman, eds. *Paleozoic systems of the Rocky Mountain Region*. Denver: Rocky Mountain Section SEPM.

Fillmore, R. 2000. *The Geology of the parks, monuments, and wildlands of southern Utah*. Salt Lake City: University of Utah Press.

Goldstrand, P. M. 1994. Tectonic development of Upper Cretaceous to Eocene strata of southwestern Utah, in *Bulletin 106*. Boulder: Geological Society of America.

Havholm, K. G., R. C. Blakey, M. Capps, L. S. Jones, D. D. King, and G. Kocurek. 1993. Eolian genetic stratigraphy: An example from the Middle Jurassic Page Sandstone, Colorado Plateau, in Pye, K., and N. Lancaster, eds. *Aeolian Sediments, Ancient and Modern*. Oxford: International Association of Sedimentologists.

Hill, C. A., W. D. Ranney, R. B. Scarborough, and J. D. Powell. 2005. Search for the ancestral

Colorado River. *Geological Society of America Abstracts with Programs* 37:36.

Hill, C. A., N. Eberz, and R. H. Buecher. 2008. A karst connection model for Grand Canyon, Arizona, USA. *Geomorphology* 95:19.

Hill, C. A., and W. D. Ranney. 2007. A proposed Laramide proto–Grand Canyon. *Geological Society of America Abstracts with Programs* 39:43.

Hintze, L. F. 1988. *Geologic history of Utah*. Provo: Brigham Young University Geologic Studies.

Holm, R. F. 2001. Cenozoic paleogeography of the central Mogollon Rim, southern Colorado Plateau region, Arizona, revealed by Tertiary gravel deposits, Oligocene to Pleistocene lava flows, and incised streams. *Bulletin 113*. Boulder: Geologic Society of America.

Hopkins, R. L. 2002. *Hiking the Southwest's Geology*: Seattle: The Mountaineers Books.

Imlay, R. 1980. Jurassic paleobiogeography of the conterminous United States in its continental setting. *Professional paper 1062*. Washington, D.C.: U.S. Geological Survey.

Jenny, J. P., and S. J. Reynolds. 1989. *Geologic evolution of Arizona*. Tucson: Arizona Geological Society.

Johnson, S. Y., M. A. Chan, and E. A. Konopka. 1992. Pennsylvanian and Early Permian paleogeography of the Uinta–Piceance basin region, northwestern Colorado and northeastern Utah. *Bulletin 1787-CC*. Washington, D.C.: U.S. Geological Survey.

Kamola, D. L., and M. A. Chan. 1988. Coastal dune facies, Permian Cutler Formation (White Rim Sandstone) Capitol Reef National Park area, southern Utah. *Sedimentary Geology* 56:341–356.

Kluth, C. F., and P. F. Coney. 1981. Plate tectonics of the ancestral Rocky Mountains. *Geology* 9:10–15.

Lageson, D. R., and J. G. Schmitt. 1994. The Sevier orogenic belt of the western United States, in Caputo, M. V., J. A. Peterson, and J. J. Franczyk, eds. *Mesozoic systems of the Rocky Mountain region, USA*. Denver: Rocky Mountain Section SEPM.

Langford, R. P., and M. A. Chan. 1989. Fluvial-aeolian interactions: Part II, ancient systems. *Sedimentology* 36:1037–1051.

Lawton, T. F. 1994. Tectonic setting of Mesozoic sedimentary basins, Rocky Mountain region, United States, in Caputo, M. V., J. A. Peterson, and K. J. Franczyk, eds. *Mesozoic systems of the Rocky Mountain region, USA*. Denver: Rocky Mountain Section SEPM.

Loope, D. B. 1984. Eolian origin of Upper Paleozoic sandstones, southeastern Utah. *Journal of Sedimentary Petrology* 54:563–580.

Loope, D. B. 1985. Episodic deposition and preservation of eolian sands: A late Paleozoic example from southeastern Utah. *Geology* 13:73–76.

Lucchitta, I. 1972. Early history of the Colorado River in the Basin and Range Province, in *Bulletin 83*. Boulder: Geological Society of America.

Lucchitta, I. 2001. *Hiking Arizona's Geology*. Seattle: The Mountaineers Books.

Luttrell, P. R. 1993. Basinwide sedimentation and the continuum of paleoflow in an ancient river system: Kayenta Formation (Lower Jurassic), central portion Colorado Plateau. *Sedimentary Geology* 85:411–434.

McKee, E. D. 1982. *The Supai Group of Grand Canyon*. Washington, D.C.: U.S. Geological Survey.

Mack, G. H. 1977. Depositional environments of the Cutler–Cedar Mesa facies transition (Permian) near Moab, Utah. *The Mountain Geologist* 14:53-68.

Marzolf, J. E. 1990. Reconstruction of extensionally dismembered early Mesozoic sedimentary basins, southwestern Colorado Plateau to eastern Mojave Desert, in Wernicke, B. P., ed. *Basin and Range extensional tectonics near the latitude of Las Vegas, Nevada*. Boulder: Geological Society of America.

Middleton, L. T., and R. C. Blakey. 1983. Sedimentologic processes and controls on the intertonguing of the fluvial Kayenta and eolian Navajo Sandstone, northern Arizona and southern Utah, in Brookfield, M. E., and T. S. Ahlbrandt, eds. *Eolian sediments and processes: Developments in sedimentology*. Amsterdam: Elsevier.

Nations, J. D. and E. Stump. 1996. *Geology of Arizona*. Dubuque: Kendall/Hunt Publishing.

Parrish, J. T., and F. Peterson. 1988. Wind directions predicted from global circulation models and wind directions determined from eolian sandstones of the Colorado Plateau, a comparison. *Sedimentary Geology* 56:261–282.

Parrish, J. T., and H. J. Falcon-Lang. 2007. Coniferous trees associated with interdune deposits in the Jurassic Navajo Sandstone formation, Utah, USA. *Paleontology* 50:829–843.

Pederson, J., K. Karlstrom, W. Sharp, and W. McIntosh. 2002. Differential incision of the Grand Canyon related to Quaternary faulting: Constraints from U-series and AR/AR dating. *Geology* 30:739–742.

Pederson, J., R. D. Mackley, and J. L. Eddleman. 2002. Colorado Plateau uplift and erosion evaluated using GIS. *GSA Today* 12:4–10.

Peterson, Fred. 1984. Fluvial sedimentation on a quivering craton: Influence of slight crustal movements on fluvial processes, Upper Jurassic Morrison Formation, western Colorado Plateau. *Sedimentary Geology* 38:21–49.

Peterson, Fred. 1988. Stratigraphy and nomenclature of Middle and Upper Jurassic rocks, western Colorado Plateau, Utah and Arizona. Washington, D.C.: U.S. Geological Survey.

Peterson, Fred, and G. N. Pipiringos. 1979. *Stratigraphic relations of the Navajo Sandstone to Middle Jurassic formations, southern Utah and northern Arizona*. Washington, D.C.: U.S. Geological Survey.

Pipiringos, G. N., and R. B. O'Sullivan.1978. *Principle unconformities in Triassic and Jurassic rocks, western interior U.S.: A preliminary report*. Washington, D.C.: U.S. Geological Survey.

Ranney, W. 2005. *Carving Grand Canyon*. Grand Canyon: Grand Canyon Association.

Rawson, R. R., and C. E. Turner-Peterson. 1980. Paleogeography of northern Arizona during the deposition of the Permian Toroweap Formation, in Fouch, T. D., and R. Magathan, eds. *Paleozoic paleogeography of the west-central United States*. Denver: Rocky Mountain Section SEPM.

Reynolds, S. J., J. E. Spencer, Y. Asmerom, E. DeWitt, and S. E. Laubach. 1989. Early Mesozoic uplift in west-central Arizona and southeastern California. *Geology* 17: 207–211.

Riggs, N. R., and R. C. Blakey. 1993. Early and Middle Jurassic paleogeography and volcanology of Arizona and adjacent areas, in Dunne, G., and K. A. McDougall, eds. *Mesozoic Paleogeography of the Western United States – II*. Los Angeles: Pacific Section SEPM.

Speed, R. C. 1978. Paleogeographic and plate tectonic evolution of the early Mesozoic marine province of the western Great Basin, in Howell, D. G., and K. A. McDougall, eds. *Mesozoic paleogeography of the western United States*. Los Angeles: Pacific Section SEPM.

Sprinkle, D. A., T. C. Chidsey, and P. B. Anderson, eds. 2000. *Geology of Utah's parks and monuments*. Salt Lake City: Utah Geological Society.

Stanesco, J. D. 1991. Sedimentology and cyclicity in the Lower Permian De Chelly Sandstone on the Defiance Plateau, eastern Arizona. *The Mountain Geologist* 28:1–11.

Stewart, J. H., F. G. Poole, and R. F. Wilson. 1972. *Stratigraphy and origin of the Chinle Formation and related Upper Triassic strata in the Colorado Plateau region*. Washington, D.C.: U.S. Geological Survey.

INDEX